SpringerBriefs in Cognitive Computation

Volume 4

Editor-in-chief

Amir Hussain, Stirling, UK

About the Series

SpringerBriefs in Cognitive Computation are an exciting new series of slim high-quality publications of cutting-edge research and practical applications covering the whole spectrum of multi-disciplinary fields encompassed by the emerging discipline of Cognitive Computation. The Series aims to bridge the existing gap between life sciences, social sciences, engineering, physical and mathematical sciences, and humanities.

The broad scope of Cognitive Computation covers basic and applied work involving bio-inspired computational, theoretical, experimental and integrative accounts of all aspects of natural and artificial cognitive systems, including: perception, action, attention, learning and memory, decision making, language processing, communication, reasoning, problem solving, and consciousness.

More information about this series at http://www.springer.com/series/10374

Raoul Biagioni

The SenticNet Sentiment Lexicon: Exploring Semantic Richness in Multi-Word Concepts

 Springer

Raoul Biagioni
Dublin Institute of Technology
Dublin 6
Ireland

ISSN 2212-6023   ISSN 2212-6031   (electronic)
SpringerBriefs in Cognitive Computation
ISBN 978-3-319-38970-7   ISBN 978-3-319-38971-4   (eBook)
DOI 10.1007/978-3-319-38971-4

Library of Congress Control Number: 2016940114

Printed on acid-free paper

This Springer imprint is published by Springer Nature
The registered company is Springer International Publishing AG Switzerland

Contents

1 Introduction.. 1
  1.1 Sentiment in Opinionated Text............................... 1
  1.2 Background 3
  1.3 Research Problem.. 5
  References.. 6

2 Sentiment Analysis 7
  2.1 Sentiment Analysis Challenges 8
  2.2 Levels of Analysis 10
  2.3 Supervised Versus Unsupervised Sentiment Analysis 10
  2.4 Linguistics-Based Sentiment Analysis........................ 11
  2.5 Lexicon-Based Sentiment Analysis 12
  2.6 Conclusion 14
  References.. 15

3 SenticNet 17
  3.1 The Common Sense Nature of SenticNet Knowledge 18
  3.2 A Seminal Approach to Concept-Based Sentiment Analysis....... 19
  3.3 Producing SenticNet.. 20
  3.4 SenticNet Processes 23
  3.5 SenticNet Knowledge: Encoding 24
  3.6 SenticNet Access Methods.................................. 25
  3.7 SenticNet in Numbers 26
    3.7.1 Concept Types: Number of Words 26
    3.7.2 Analysis of Polarity Values: Single-Word
            Versus Multi-word Concepts 27
    3.7.3 Most Common Part-of-Speech Classes 29
  3.8 Conclusion 30
  References.. 30

4 Unsupervised Sentiment Classification 33
  4.1 Datasets 33

    4.2   Classification Design and Implementation.   34
          4.2.1   Overview   34
          4.2.2   Sentiment Classification Process   35
          4.2.3   Polarity Value Thresholds.   35
          4.2.4   Implementation.   38
    4.3   Conclusion   42
    References.   42

5   Evaluation   45
    5.1   Classification Performance.   45
    5.2   Research Question   47
    5.3   Qualitative Differences Between the Datasets   47
    5.4   SenticNet   48
    5.5   Sentiment Analysis System   49
    5.6   Sentiment Classification Design.   49
    5.7   Limitations   49
    5.8   Conclusions   50

6   Conclusion   51
    6.1   Conclusion   51
    6.2   Future Work   52
    6.3   Final Remarks   52
    Reference.   53

Index.   55

Chapter 1
Introduction

Abstract The theme of this paper has its origins in linguistic theory, namely that the meaning of multi-word concepts such as "ice cream", "acid rain" or "city wall" cannot be understood simply from the meaning of their individual parts. The semantically richer nature of multi-word concepts that ensues from their specificity is the subject of analysis of this paper. This chapter contains three main sections. Section 1.1 discusses the ubiquitous nature of opinionated digital text, and the potential application domains for the analysis of sentiment expressed in such texts. Section 1.2 provides an overview of existing sentiment analysis approaches and introduces the SenticNet sentiment lexicon. Section 1.3 presents the research question this paper aims to answer, namely whether multi-word concepts outperform single-word concepts in a sentiment classification task by virtue of their semantic richness. The section outlines the reasons why SenticNet was deemed an ideal platform for investigating the research question.

Keywords Concepts · Polarity · Semantic richness · Sentiment analysis · Sentiment lexicon · SenticNet

1.1 Sentiment in Opinionated Text

People express opinions online every day concerning trivial or important issues, whether the topic is politics, fashion or films, a hotel or a restaurant, a flat screen TV or a baby carrier. Each day, 154 billion e-mails, 500 million tweets, 1 million blog posts and 1.3 million WordPress blog comments are written. In the United States alone, Facebook users write approximately 16 billion words per day, while e-mail and utterances via social media roughly equate to 3.6 trillion words daily, or the equivalent of 36 million books every day (Harris 2014). This vast reservoir of digital texts is, in many cases, opinionated by virtue of the platforms from which the texts originate: social networks, review sites, customer feedback emails, personal blog pages, forums, tweets, and so on.

© The Author(s) 2016
R. Biagioni, *The SenticNet Sentiment Lexicon: Exploring Semantic Richness in Multi-Word Concepts*, SpringerBriefs in Cognitive Computation 4,
DOI 10.1007/978-3-319-38971-4_1

The opinions of others are a valuable source of information for making decisions. For individuals, the opinions of peers can help in making choices, especially when decisions involve multiple options from which to choose. For organisations, the opinions of customers can assist in gaining and sustaining competitive advantage. Individuals wish to learn from others' past experiences before making purchasing decisions. Organisations want to know how their products are perceived, how they can improve them or how to approach the development of new products. Collective likes and dislikes with regard to products, personalities and issues are nowadays easily accessible in abundance online. While, in the past, organisations or businesses had to conduct surveys, opinion polls, or focus groups in order to gather public or consumer opinions (Liu 2012), they now have the possibility of leveraging sentiment expressed in opinionated text through automated means. The automated processing of opinions and the sentiment expressed therein is referred to as sentiment analysis.

Application domains for sentiment analysis include online customer reviews, marketing and business intelligence, and social networks. Typically, sentiment analysis systems process articles, news items, tweets or social network posts from various sources that discuss or mention a given organisation or issue, in order to aggregate the sentiment expressed in them. As indicated in the examples below, these sources provide fertile ground for the application of sentiment analysis techniques:

- Numerous websites offer users a forum through which they can provide feedback and reviews of products. From a prospective customer point of view, reading hundreds of reviews is a very time-consuming task. In addition, many reviews are often conflicting, making it difficult for readers to find value in such reviews. These large quantities of online opinions lend themselves to being processed by sentiment analysis applications (Duric and Song 2012). One such application is Google Shopping, which provides online sentiment summaries of reviews about products and their properties. Google's sentiment analysis system collects product reviews from third party and seller websites, and subsequently categorises them by product properties, such as "service" or "value", summarising sentiment information regarding those aspects (Feldman 2013).
- Similar to the situation faced by prospective customers mentioned above, spending a great deal of time sifting through customer reviews of their products, companies can use sentiment analysis to aggregate and summarise reviews into information that can help them learn about the relationship between their customers and their products (Duric and Song 2012). An example of sentiment analysis applications that can help organisations achieve this goal are social media monitoring applications. Such applications are used to monitor the reputation of a brand on Twitter and/or Facebook (Feldman 2013). For example, Trackur allows a user to perform a keyword search (e.g. a brand name) that returns relevant posts sourced from news sites, Facebook, Twitter, etc. The search results are displayed via colour-coded sentiment flags, to indicate whether a post is positive or negative.
- Sentiment analysis can support organisations in marketing their products via contextual online advertising. Contextual advertising refers to the practice of targeting products only to consumers who are likely to be receptive to the organisation's offering (Duric and Song 2012). Contextual advertising systems

ensure that, for example, product advertisements are not displayed on blog pages in which the review of that particular product is negative. Similarly, if a blog mentions a product favourably, a purchase decision is more likely, and so the product can be advertised on that page. An example of this approach is the blog website Tumblr (Seward 2013).

- Sentiment analysis systems could also be applied to the issue of cyber-bullying. This refers to the use of electronic communication to bully a person, typically by sending messages of an intimidating or threatening nature. As social media websites have become popular among children and teenagers, the issue of cyber-bullying has emerged as a problem that requires serious attention, both from parents and social media platform owners. Sentiment analysis techniques could provide valuable methods to help combat this recent phenomenon because, as with opinions about products, they can readily deal with opinions about other people (Duric and Song 2012).

- Sentiment analysis techniques could also potentially be integrated into a general-purpose "sentiment search engine" (Duric and Song 2012). Currently, when searching the Internet using a commonly used search engine such as Google or Bing, the search results do not specifically incorporate a sentiment dimension in search results. A sentiment search engine could rank results according to polarities, or generate summaries or scores based on the sentiment information in the websites that are returned by the search.

1.2 Background

Sentiment analysis, also called opinion mining, is the field of study that analyses people's opinions, sentiments, evaluations, appraisals, attitudes, and emotions towards entities such as products, services, organizations, individuals, issues, events, topics, and their attributes (Liu 2012, p. 1).

The field of study at the core of this paper is sentiment analysis. The above definition of "sentiment analysis" suggests that the terms "opinion" and "sentiment" are interchangeable. Similarly, Pang and Lee (2008), and Medhat et al. (2014) use the terms "sentiment analysis" and "opinion mining" interchangeably. Pang and Lee state that, if "broad interpretations are applied, sentiment analysis and opinion mining denote the same field of study". Medhat et al., meanwhile, state that "The two expressions [...] are interchangeable. They express a mutual meaning". Cambria et al. (2010a), conversely, provide distinct definitions for each term. They view sentiment analysis as "inferring emotional states from text", and opinion mining as being concerned with "the identification of opinions and attitudes in natural language texts". These divergent views of the exact meaning of sentiment analysis prompted the need to set the boundaries for this paper, with regard to terminology, from the outset. Thus, in this paper, the definition of "sentiment" given by the Oxford Dictionaries online[1]—"a view or opinion that is held or expressed"—is adopted.

---

[1]http://www.oxforddictionaries.com/definition/english/sentiment.

Throughout this paper terms such as "sentiment", "opinion", "affect" and "emotion" are to be conceived as having either positive or negative "sentiment polarity". The term "sentiment classification" refers to the task of determining whether the sentiment in a text is positive or negative. A variant of this, not covered by this paper, is to determine whether a text is factual (i.e. objective or neutral) or opinionated (i.e. subjective).

Sentiment analysis can be performed using different strategies, depending on the type of information considered to be most effective for the task at hand. Information sources for sentiment analysis can be of a linguistic, statistical, similarity or implicit nature. Linguistic information can be gathered through means such as affiliation to a part-of-speech class, syntactic sentence structure or lexical relationships (e.g. synonyms). A part-of-speech class is a lexical category which indicates the role a word plays in the sentence in which it appears. Part-of-speech information provides information about the semantic content of a word. For example, nouns usually denote "tangible and intangible things," whereas prepositions express relationships between "things" (Feldman and Sanger 2006). Syntactic sentence structure refers to the way words are joined together (Burling 1992). One of the most widely used statistical sources of information is term frequency i.e. individual words (unigrams) or multiple words (bigrams, trigrams, n-grams) and their frequency counts. Similarity information can be gathered through means such as affiliation to a given sentiment polarity. Implicit sentiment information is suggested but not directly expressed. Such information is obtained from the sentiment underpinning the concepts that comprise a text. Sentiment analysis strategies that rely on implicit sentiment information are therefore also referred to as "concept-based sentiment analysis". A "concept" is to be understood as defined by Merriam-Webster dictionary online[2]—"an abstract or generic idea generalized from particular instances"—and can take the form of a single word e.g. "love" or of a multi-word expression e.g. "accomplish goal". Examples of concepts that carry implicit sentiment information are "birthday cake", "accomplish goal" and "lose temper".

Concept-level sentiment analysis is a novel approach to the analysis of opinionated text conceived at the MIT Media Laboratory in 2010 (Cambria et al. 2010a). Since 2014, the research has been further developed at NTU School of Computer Science in Singapore. Concept-level sentiment analysis focuses on the implicit sentiment associated with concepts rather than statistical, linguistic, or similarity information (Cambria and White 2014) and thereby introduces semantics, or meaning, into the sentiment analysis process.

The research has produced a publicly available sentiment lexicon, named SenticNet (Cambria et al. 2014). SenticNet is a knowledge-based approach to concept-level sentiment analysis (Cambria et al. 2013). The purpose of SenticNet is to act as a resource for sentiment analysis tasks. The premise of SenticNet is that it is possible to make the conceptual and sentiment information conveyed by natural language accessible to computers. The sentiment lexicon SenticNet consists of a

---

[2]http://www.merriam-webster.com/dictionary/concept.

collection of almost 14,000 affective concepts; i.e. concepts that carry sentiment information, and their corresponding sentiment polarity value. SenticNet is the first sentiment lexicon to capture multi-word concepts, alongside single-word concepts, in a systematic fashion.

The theme of this paper has its origins in linguistic theory, namely that the meaning of two-word sequences such as "ice cream", "acid rain" or "city wall" cannot be understood simply from the meaning of their individual parts. "Ice cream", for example, has a more specific sense than what could be understood from the meanings of "ice" and "cream" (Burling 1992). This conception of two-word sequences is extended in this paper to multi-word concepts such as "feel happy" where the concept is seen as more specific than its constituent parts "feel" and "happy". The semantically richer nature of multi-word concepts that ensues from their specificity is the subject of analysis of this paper. By storing both single-word and multi-word concepts, SenticNet is an ideal platform for investigating the research question posed in the next section.

1.3   Research Problem

* Do multi-word concepts, by virtue of their semantic richness, outperform single-word concepts in a sentiment classification task?

In SenticNet, sentiment information is distributed across two types of concepts: single-word concepts and multi-word concepts. The research problem addressed in this paper is the context sensitivity of single-word concepts. For example, the word "unpredictable" may have a negative sentiment in a car review (as in "unpredictable steering"), but it could have a positive sentiment in a movie review (as in "unpredictable plot") (Liu 2012). Multi-word concepts, conversely, allow one to mitigate this issue, because, by virtue of their semantic richness, they convey context information that single-word concepts cannot. SenticNet could potentially be more reliable, and thus could also be of greater value, if it would not contain the inherently ambiguous single-word concept type, and instead would solely provide sentiment information for the more semantically rich multi-word concept type.

The proposition that SenticNet could gain from being solely a multi-word concept sentiment lexicon prompted the research question of whether multi-word concepts, by virtue of their semantic richness, outperform single-word concepts in a sentiment classification task. This research question is of interest because, if multi-word concepts could be shown to be better sentiment indicators than single-word concepts, this insight could feed into future decisions about the shape and content of the SenticNet sentiment lexicon. Furthermore, any findings with regard to the differences between single- and multi-word concepts in a sentiment analysis context could contribute to the research field of concept-based sentiment analysis as a whole. Prior to the introduction of SenticNet it was not possible to investigate this research question, because traditional sentiment lexicons consisted of mainly single-word concepts.

References

Burling R (1992) Patterns of language: structure, variation, change, 1st edn. Academic Press Inc., San Diego, California

Cambria E, White B (2014) Jumping NLP curves: a review of natural language processing research [Review Article]. IEEE Comput Intell Mag 9:48–57. doi:10.1109/MCI.2014.2307227

Cambria E, Hussain A, Havasi C, Eckl C (2010a) Sentic computing: exploitation of common sense for the development of emotion-sensitive systems. In: Proceedings of the second international conference on development of multimodal interfaces: active listening and synchrony, COST'09. Springer, Berlin, Heidelberg, pp 148–156. doi:10.1007/978-3-642-12397-9_12

Cambria E, Schuller B, Liu B, Wang H, Havasi C (2013) Knowledge-based approaches to concept-level sentiment analysis. IEEE Intell Syst 12–14

Cambria E, Olsher D, Rajagopal D (2014) SenticNet 3: a common and common-sense knowledge base for cognition-driven sentiment analysis. In: Twenty-eighth AAAI conference on artificial intelligence

Duric A, Song F (2012) Feature selection for sentiment analysis based on content and syntax models. Decis Support Syst 53:704–711. doi:10.1016/j.dss.2012.05.023

Feldman R (2013) Techniques and applications for sentiment analysis. Commun ACM 56:82. doi:10.1145/2436256.2436274

Feldman R, Sanger J (2006) The text mining handbook advanced approaches in analyzing unstructured data. Knowledge management, databases and data mining

Harris J (2014) Sturgeon's law, big data and the new literacy. Inf Manag

Liu B (2012) Sentiment analysis and opinion mining. Synth Lect Hum Lang Technol 5:1–167. doi:10.2200/S00416ED1V01Y201204HLT016

Medhat W, Hassan A, Korashy H (2014) Sentiment analysis algorithms and applications: a survey. Ain Shams Eng J. doi:10.1016/j.asej.2014.04.011

Pang B, Lee L (2008) Opinion mining and sentiment analysis

Seward ZM (2013) How Yahoo plans to make money on Tumblr: ads that don't feel like ads [WWW Document]. Yahoo Finance. http://finance.yahoo.com/news/yahoo-plans-money-tumblr-ads-211528425.html. Accessed 28 Jan 15

Chapter 2
Sentiment Analysis

Abstract There can, in principle, be no systematic discovery procedure that would yield a set of grammatical rules. The discovery of the rules of language is like the discovery of the rules of nature, a stumbling and unsystematic affair that calls primarily for insight and imagination (Chomsky cited in Burling, Patterns of language: structure, variation, change, 1992, p. 56). This chapter presents a review of the literature in the research field of sentiment analysis to provide an overview of the sentiment lexicon landscape in which to situate SenticNet. The chapter contrasts the concept-based approach to sentiment analysis to linguistics-based approaches, in order to elicit a better understanding of the potential advantages of the concept-based model over the linguistic-based model. The choice of contrasting these two approaches is motivated by their shared interest in integrating semantics into the solution to the problem of detecting sentiment polarity. The near impossibility of producing a complete set of linguistic rules, as suggested by Chomsky, on which to model the detection of sentiment polarity lays the foundations for building the case for the concept-based approach enabled by SenticNet, and further supports the interest in the research question posed in Sect. 1.3. The chapter consists of five main sections. Section 2.1 provides an overview of typical challenges encountered in sentiment analysis. Section 2.2 discusses the levels at which sentiment analysis can be performed. Section 2.3 discusses supervised and unsupervised learning approaches to sentiment analysis. Section 2.4 discusses linguistics-based sentiment analysis. Section 2.5 discusses lexicon-based sentiment analysis.

Keywords Affective knowledge · Linguistics · Machine learning · Part-of-speech · Semantic orientation · Syntactic patterns · Synsets

The goal of this chapter is to contrast the concept-based approach to sentiment analysis to linguistics-based approaches, in order to elicit a better understanding of the potential advantages of the concept-based model over the linguistic-based model. The choice of contrasting these two approaches is motivated by

© The Author(s) 2016
R. Biagioni, *The SenticNet Sentiment Lexicon: Exploring Semantic Richness in Multi-Word Concepts*, SpringerBriefs in Cognitive Computation 4,
DOI 10.1007/978-3-319-38971-4_2

their shared interest in integrating semantics into the solution to the problem of detecting sentiment polarity. The near impossibility of producing a complete set of linguistic rules, as suggested by Chomsky, on which to model the detection of sentiment polarity lays the foundations for building the case for the concept-based approach enabled by SenticNet, and further supports the interest in the research question posed in Sect. 1.3. In addition, this chapter aims to provide an overview of typical challenges encountered in sentiment analysis and of the sentiment lexicon landscape in which to situate SenticNet.

This chapter presents a review of the literature in the research field of sentiment analysis. The chapter consists of five main sections. Section 2.1 discusses sentiment analysis challenges. Section 2.2 discusses the levels at which sentiment analysis can be performed. Section 2.3 discusses supervised and unsupervised learning approaches to sentiment analysis. Section 2.4 discusses linguistics-based sentiment analysis. Section 2.5 discusses lexicon-based sentiment analysis.

2.1 Sentiment Analysis Challenges

The review of relevant research literature in the field of study of sentiment analysis highlighted the most commonly encountered challenges as being related to the issues of context sensitivity; negation; sarcasm, irony and idioms; the intensity of sentiment; implicit sentiment; compound sentences; ambiguity and domain sensitivity. Each of these challenges is further outlined below.

Context
The sentiment polarity of a word often depends on its context (Ding et al. 2008). For example the word mind is positive in "to be in the right mind" but negative in "to lose one's mind". Other words have a lack of sensitivity to context. For example, a word such as "excellent" is positive in almost all contexts.

Negation
Including detection of negation in automated sentiment analysis is of critical importance because negation terms such as "not" change the polarity of opinion words (Liu 2012). For example, in the sentence "the picture quality is not good", the polarity of the adjective "good" is inverted by the negation term "not". Negation-handling is not trivial, as not all appearances of explicit negation terms reverse the polarity of the enclosing sentence e.g. "No wonder this is considered one of the best" (Pang and Lee 2008).

Sarcasm, irony and Idioms
Negation can often be expressed in rather subtle ways through sarcasm and irony, which are quite difficult to detect. Such sentences become particularly challenging for sentiment analysis systems when they include sentiment words as illustrated by the sentences "Some cause happiness wherever they go; others whenever they go", "What a great car! It stopped working in two days" (Pang and Lee 2008; Liu 2012).

Similarly, while most idioms (e.g. "cost someone an arm and a leg") express strong opinions, they are challenging to deal with because of the absence of sentiment words (Ding et al. 2008).

Intensifiers
Intensifiers are terms that change the degree of sentiment (Kennedy and Inkpen 2006). For example, in the sentence "This movie is very good", the terms "very good" have stronger positive polarity than the single word "good" alone. On the other hand, in the sentence "This movie is barely any good", the diminisher "barely" makes the statement less positive.

Implicit opinions and sentiment
Many sentences imply opinions or sentiment without resorting to the use of sentiment words (Hu and Liu 2004). For example, in the sentence "While light, it will not easily fit in pockets", the size of the camera is the opinion target, but the word "size" is not explicitly mentioned in the sentence.

Compound sentences
Compound sentences may express more than one opinion. For example, the sentence, "The picture quality of this camera is amazing and so is the battery life, but the viewfinder is too small for such a great camera", expresses both positive and negative opinions (one may say that it has a mixed opinion). For "picture quality" and "battery life", the sentence is positive, but for "viewfinder", it is negative. It is also positive for the camera as a whole.

Ambiguity
A significant challenge in natural language processing is how to handle the ambiguity that arises when interpreting a single sentence. Lexical ambiguity is where different possible meanings are associated with one word, depending on its part-of-speech class. A part-of-speech class is a lexical category such as noun or adjective. For example, the meaning of the word "back" changes depending on whether it is an adverb (e.g. "go back"), an adjective (e.g. "back door"), a noun (e.g. "the back of the room") or a verb (e.g. "back up your files"). Syntactic ambiguity refers to a situation where a sentence may be interpreted in more than one way due to ambiguous sentence structure. For example, "I saw the man with the binoculars" has two interpretations: one where the prepositional phrase "with the binoculars" relates to the noun "man" and one where it relates to the verb "saw". Syntactic ambiguity leads to semantic ambiguity, because "one interpretation means that the man has binoculars and the other means that I used binoculars" (Russell and Norvig 2009). While humans can comprehend the intended meanings by inferring from context or by drawing on personal knowledge and understanding, computers don't benefit from the subtleties of human experience and learning. Resolving ambiguity is one of the most difficult tasks in natural language processing because computers lack "common sense" (PARC Natural Language Processing 2007).

Domain Sensitivity
A key problem in sentiment analysis is its sensitivity to the domain from which either training data is sourced, or on which a sentiment lexicon is built. This is

referred to as the problem of domain dependence. A domain refers to the topic of a given text; e.g. cars or movies. The same word in one domain may have positive polarity, while, in another domain, it has negative polarity. For example "unexpected" has positive polarity in the movie domain (e.g. "unexpected plot") but negative polarity in the automotive domain (e.g. "unexpected steering") (Liu 2012). Machine learning methods are highly sensitive to the domain from which the training data has been extracted, and can not be employed for domains in which no training data exists. A classifier trained using opinionated texts from one domain often performs poorly when it is applied to opinionated texts from another domain. With regard to lexical-based methods it is a challenging task to compile general-purpose sentiment lexicons that are capable of performing well for every topic.

2.2 Levels of Analysis

Sentiment analysis can be performed at different levels: document-level, sentence-level, entity-level and aspect-level. Document-level sentiment analysis classifies a whole opinionated document as either positive or negative (Pang et al. 2002; Turney 2002; Godbole et al. 2007; Devitt and Ahmad 2007). Liu (2012) notes that this level of analysis assumes that each document expresses opinions on a single entity (e.g. a single product), and thus is not applicable to documents which evaluate or compare multiple entities. Sentence-level sentiment analysis classifies individual sentences as positive or negative (Bethard et al. 2004; Kim and Hovy 2004). Liu (2010) notes that an assumption of sentence-level sentiment classification is that sentences express a single opinion from a single opinion-holder; e.g. "The picture quality of this camera is amazing". However, as noted in the previous section, compound sentences may express more than one opinion. Entity-level and aspect-level sentiment analysis is based on the idea that sentiments have a target. For example, in the sentence "Although the service is not that great, I still love this restaurant", the sentence adopts a positive stance concerning the entity "restaurant" but a negative stance regarding the aspect of "service". Thus, the goal of this level of analysis is to discover sentiments on entities and/or their features (Yi et al. 2003; Sebastiani et al. 2006).

2.3 Supervised Versus Unsupervised Sentiment Analysis

Sentiment analysis can be performed in a supervised or unsupervised manner. The term "supervised" refers to a methodology whereby applications learn to make classification decisions from labelled examples in training data. Labelled training data refers to data that has been manually annotated, that is, whether a text or sentence expresses positive or negative sentiment. Such applications require large amounts of domain-specific training data in order to learn effectively. Creating labelled training data for all possible domains is expensive and time-consuming. Supervised sentiment analysis is therefore limited by the availability of domain-specific labelled data. The term "unsupervised" refers to a methodology whereby applications classify data without using training data or creating models. Unsupervised sentiment analysis, conversely, does not require labelled training data, and is domain-independent. Typically, unsupervised learning occurs through the clustering, or grouping, of documents or sentences belonging to the same positive or negative sentiment polarity category.

Supervised sentiment analysis approaches are based on finding a suitable set of classification features. Classification features are distinctive attributes in a given text that are suitable for effectively discriminating between positive and negative sentiment. Such features were introduced in Sect. 1.2 as "sources of information" on which sentiment analysis strategies can be based. Taboada et al. (2011) found that supervised methods achieve high accuracy in detecting sentiment polarity in natural language text. Supervised sentiment analysis techniques commonly use the machine learning algorithms Naive Bayes classifier, Maximum Entropy and Support Vector Machine (SVM). These algorithms are described in detail in Alpaydin (2010). Unsupervised sentiment analysis approaches are based on determining the semantic orientation of specific words or phrases within a sentence or document to infer their sentiment orientation. A phrase is a small group of words standing together as a conceptual unit; e.g. "to improve standards". The two main unsupervised approaches to sentiment analysis are linguistics-based (e.g. Turney 2002) or lexicon-based (e.g. Taboada et al. 2011; Feldman 2013).

2.4 Linguistics-Based Sentiment Analysis

To computers text is a simple sequence of character strings. This makes it challenging to automatically extract the most useful elements contained in text. Linguistics-based sentiment analysis divides text and sentences not only into their constituent words but also identifies their syntactic structure.

Sentiment analysis based on linguistic techniques aims to find part-of-speech patterns that are most likely to express an opinion. Such patterns can take the form of word sequences such as an adjective followed by a noun, or the form of more complex syntactic categories such as noun or verb phrases. These, in turn, can be combined into "trees" representing the nested phrase structure of sentences (Russell and Norvig 2009).

Liu (2012) notes that adjectives have been shown to be important indicators of opinions. This is supported by the observation that several researchers have used adjectives as their primary sentiment detection feature. A brief description of past sentiment analysis research based on adjectives and/or part-of-speech patterns is given below:

- Hatzivassiloglou and McKeown (1997) use a list of adjectives, manually labelled as positive or negative, to correlate the conjunctions "but" or "and" with semantic orientation. Conjunctions are linguistic features that connect words, sentences, phrases or clauses. Hatzivassiloglou and McKeown demonstrated that the conjunctions provide indirect information about sentiment orientation of conjoined words i.e. that they have the same sentiment polarity. For example, the adjectives "fair and legitimate" and "corrupt and brutal" are in each case of identical polarity.
- Turney (2002) classifies sentiment of customer reviews sampled from different domains (automobiles, banks, movies, and travel destinations) using rules based on heuristic part-of-speech patterns that are considered likely to be used to express opinions. The patterns consist of adjectives (JJ), adverbs (RB, RBR, RBS), verbs (VB, VBD, VBN, VBG) and nouns (NN, NNS). This approach is based on the idea that although adjectives often indicate subjectivity, they provide insufficient context to reliably determine sentiment polarity as illustrated in Sect. 1.3.
- Benamara et al. (2007) perform part-of-speech-based sentiment analysis using adverbs and adverb-adjective combinations e.g. "very bad". This approach is based on the observation that adverbs usually act as modifiers of adjectives. A modifier is defined in the Oxford Dictionaries online[1] as "a word, especially an adjective or noun [...] that restricts or adds to the sense of a head noun".
- Ding et al. (2008) use adjective, adverb, verb and noun to determine the orientation of opinions about product features in customer reviews. The inclusion of part-of-speech classes other than adjectives is based on the fact that, while orientations apply to most adjectives, there are those adjectives that have no orientations (e.g., external, digital).

2.5 Lexicon-Based Sentiment Analysis

Lexicons that contain compilations of words annotated with sentiment orientation are referred to as sentiment lexicons (Ohana 2009). The rationale underlying lexicon-based techniques is the assumption that the most important indicators of sentiment in natural language text are sentiment words, also called opinion words. These are words that are commonly used to express positive or negative sentiments. For example, "good", "wonderful" and "amazing" are positive sentiment words, while "bad", "poor" and "terrible" are negative sentiment words (Liu 2012). Sentiment lexicons relate words to sentiment polarity.

---

[1]http://www.oxforddictionaries.com/definition/english/modifier.

The Sentiment Lexicon Landscape

In this section, an overview of past and current sentiment lexicons is presented:

- Stone and Hunt (1963) developed a system for content analysis named "the General Inquirer". The system enabled the understanding of "the psychological forces" in a document. The General Inquirer system used two dictionaries: a purpose-built psycho-sociological dictionary, and an anthropological dictionary used for studying themes in the folktales of different cultures. The former contains category information such as "Persons", "Physical Objects", "Emotions" or "Culture", while the latter contains categories such as "Dominate", "Follow", "Leader", "Power" or "Agree". When analysing a sentence, GI looked up words in the dictionaries and, if a match was found, tags indicating the word's membership in one or more categories specified in the dictionaries were attached to the sentence.

- Esuli and Sebastiani (2005) used entries in the English language dictionary of synonyms WordNet (Kamps et al. 2004) to create the sentiment lexicon SentiWordNet. Starting from the lists of positive and negative words used by Turney and Littman (2003), SentiWordNet was "grown" by repeatedly searching for and adding word synonyms from WordNet. Synonyms in WordNet are grouped into "synsets". Each synset is annotated with a definition, example uses and their relationship with other synsets. Wordnet does not contain sentiment-related information. Terms in SentiWordNet may have multiple "senses" i.e. meanings. For example, the term "cold" can mean "having a low temperature" or "without human warmth or emotion". To allow differentiation between the senses of terms, SentiWordNet provides descriptions, named glosses, for each entry. The fact that terms in SentiWordNet may be stored multiple times, based on their sense, has implications for sentiment analysis applications. For example, the polarity value of the adjective "cold", with the meaning "having a low temperature" (e.g. "cold beer"), is different from the polarity value of the adjective "cold" with the meaning "being emotionless" (e.g. "cold person").

- Strapparava and Valitutti (2004) created the sentiment lexicon WordNet-Affect based on a subset of WordNet synsets. The lexicon consists of "affective knowledge". Affective knowledge is an umbrella term used for words that express moods, feelings and attitudes. In WordNet-Affect, synsets are associated with an affective label. An affective label represents an affective category such as an emotion, feeling, etc. WordNet-Affect is, in effect, a variant of WordNet where synsets have been enriched with affective meaning. For example, in WordNet-Affect, the synset "anger" is attached to the label "emotion".

SentiWordNet and WordNet-Affect are the most widely used sentiment lexicons (Poria et al. 2013). Both lexicons are mostly limited to single-word concepts (Poria et al. 2013).

Limitations of Sentiment Lexicons

Sentiment words provide important clues for sentiment analysis. However, the sole reliance on sentiment lexicons is, as described in Liu (2012) and Pang and Lee (2008), subject to potential limitations:

- Sentiment lexicons are domain-independent. This affects the ability of sentiment analysis systems that rely on such lexicons to deal with the fact that the same term may have opposing polarity, depending on the domain in which it is used. For example, "go read the book" most likely indicates positive sentiment for book reviews, but negative sentiment for movie reviews. Dealing with such situations is referred to as word sense disambiguation.
- Domain independence makes sentiment lexicons portable. Portability refers to the ability to apply sentiment lexicons on text from any domain, be it movies, digital cameras or emotion-related texts. However, the downside to portability is that it makes sentiment lexicons less precise when confronted with the disambiguation scenarios described above.
- Similarly to the above issue, words can have different polarities within the same domain. For example, in the camera domain, the word "long" clearly expresses opposite opinions in the following two sentences: "The battery life is long" (positive) and "It takes a long time to focus" (negative).
- Words found in sentiment lexicons do not always express sentiment. For example, in "I am looking for a good car to buy", "good" does not express a positive or negative sentiment.
- Traditional sentiment lexicons cannot deal with implicit (or implied) sentiment. For example, in the sentence "the food is too expensive", the concept of "price" is an implicit information source, as it does not appear in the sentence but it is implied.
- It is difficult to create unique sentiment lexicons covering all possible language constructs. For example, slang is common in online social networks but is not typically supported in sentiment lexicons.

2.6   Conclusion

The sentiment analysis literature review performed in this chapter is relevant to the research question in all aspects. The review of challenges associated with sentiment analysis highlighted the obstacles that are likely going to be faced during the implementation phase of the research and provide valuable insights into the diversity of issues that must be considered when performing sentiment analysis. The review of the levels of analysis of sentiment analysis indicated that testing whether multi-word concepts outperform single-word concepts can be done at either document-, or sentence-level under the condition that the whole document or sentence expresses opinions on a single opinion target. The sentiment analysis methodologies reviewed in Sect. 2.3 indicated that the supervised sentiment analysis approach can be employed under the condition that a suitable pre-labelled dataset is available. Furthermore, the selection of classification features as applied in the supervised approaches can be applied to both concept types. However, the unsupervised lexicon-based approach appeared as the most appropriate approach in light of the fact that the research is specifically focused towards the SenticNet sentiment lexicon.

The review of linguistics-based sentiment analysis complements the position expressed by Chomsky, that it is not possible to harness natural language grammar in an automated way. The research examples described in this chapter illustrate that linguistic rules for identifying sentiment are hand-crafted and heuristic. No universal and exhaustive sentiment grammar exists. This finding provides additional support for lexicon-based approaches in general, and concept-based approaches in particular. While sentiment lexicons are also not exhaustive, they are not restricted by limitations that are out of their control such as the complexities of a language. A lexicon can be as exhaustive as is wanted by its authors. Similarly concepts are free from restrictions imposed by part-of-speech affiliation. In the research examples described in this chapter, approaches relied on adverbs, adjectives, nouns, etc. and sequences thereof. However, combinations of nouns such as "birthday gift" were not attempted because the pattern in itself does not indicate the presence of possible sentiment information. Lastly, the review of the sentiment lexicon landscape illustrates the novel aspects of SenticNet by highlighting the lexical, word-based, nature of traditional sentiment lexicons. SenticNet, in contrast, is a concept-based lexicon in which lexical affinity does not play a role.

After having ascertained aspects of the research question from a field of study point of view, the next step is to investigate and analyse SenticNet.

References

Alpaydin E (2010) Introduction to machine learning, 2nd edn. MIT Press, Cambridge, Mass

Benamara F, Cesarano C, Picariello A, Reforgiato D, Subrahmanian VS (2007) Sentiment analysis: adjectives and adverbs are better than adjectives alone. In: Proceedings of International Conference Weblogs Social Media ICWSM

Bethard S, Yu H, Thornton A, Hatzivassiloglou V, Jurafsky D (2004) Automatic extraction of opinion propositions and their holders. In: In 2004 AAAI spring symposium on exploring attitude and affect in text, pp. 22–24

Burling R (1992) Patterns of Language: structure, variation, change, 1st edn. Academic Press Inc., San Diego, California

Devitt A, Ahmad K (2007) Sentiment polarity identification in financial news: a cohesion-based approach. In: Proceedings of the 45th Annual Meeting of the Association of Computational Linguistics

Ding X, Liu B, Yu PS (2008) A holistic lexicon-based approach to opinion mining. In: Proceedings of the 2008 international conference on web search and data mining. ACM, pp. 231–240

Esuli A, Sebastiani F (2005) Determining the semantic orientation of terms through gloss classification. In: Proceedings of the 14th ACM international conference on information and knowledge management, CIKM'05. ACM, New York, NY, USA, pp 617–624. doi:10.1145/1099554.1099713

Feldman R (2013) Techniques and applications for sentiment analysis. Commun ACM 56:82. doi:10.1145/2436256.2436274

Godbole N, Srinivasaiah M, Skiena S (2007) Large-scale sentiment analysis for news and blogs. In: Proceedings of the international conference on weblogs and social media (ICWSM)

Hatzivassiloglou V, McKeown KR (1997) Predicting the semantic orientation of adjectives. In: Proceedings of the 35th annual meeting of the association for computational linguistics and eighth conference of the European chapter of the association for computational linguistics, ACL'98. Association for Computational Linguistics, Stroudsburg, PA, USA, pp 174–181. doi:10.3115/976909.979640

Hu M, Liu B (2004) Mining opinion features in customer reviews. In: AAAI, pp 755–760

Kamps J, Marx MJ, Mokken RJ, De Rijke M (2004) Using wordnet to measure semantic orientations of adjectives

Kennedy A, Inkpen D (2006) Sentiment classification of movie reviews using contextual valence shifters. Comput. Intell 22:110–125

Kim S-O, Hovy E (2004) Determining the sentiment of opinions. Presented at the proceedings of the COLING conference, Geneva

Liu B (2010) Sentiment analysis and subjectivity. In: Handbook of natural language processing, 2nd edn. Taylor and Francis Group, Boca

Liu B (2012) Sentiment analysis and opinion mining. Synth Lect Hum Lang Technol 5:1–167. doi:10.2200/S00416ED1V01Y201204HLT016

Ohana B (2009) Opinion mining with the SentWordNet lexical resource. Dublin Institute of Technology

Pang B, Lee L (2008) Opinion mining and sentiment analysis

Pang B, Lee L, Vaithyanathan S (2002) Thumbs up? sentiment classification using machine learning techniques. In: Proceedings of the ACL-02 conference on empirical methods in natural language processing-volume 10. Association for Computational Linguistics, pp 79–86

PARC Natural Language Processing (WWW Document) (2007) http://www.parc.com/content/attachments/naturallanguage_backgrounder_parc.pdf. Accessed 16 Jan 2015

Poria S, Gelbukh A, Hussain A, Howard N, Das D, Bandyopadhyay S (2013) Enhanced senticnet with affective labels for concept-based opinion mining. IEEE Intell Syst 28:31–38. doi:10.1109/MIS.2013.4

Russell S, Norvig P (2009) Artificial intelligence: a modern approach, 3rd edn. Prentice Hall, Upper Saddle River

Sebastiani AF, Esuli A, Sebastiani F (2006) Determining term subjectivity and term orientation for opinion mining andrea Esuli. In: In Proceedings of the 11th conference of the european chapter of the association for computational linguistics (EACL'06

Stone PJ, Hunt EB (1963) A computer approach to content analysis: studies using the general inquirer system. In: Proceedings of the May 21–23, 1963, spring joint computer conference, AFIPS'63 (Spring). ACM, New York, NY, USA, pp 241–256. doi:10.1145/1461551.1461583

Strapparava C, Valitutti A (2004) WordNet affect: an affective extension of wordnet. In: LREC, pp 1083–1086

Taboada M, Brooke J, Tofiloski M, Voll K, Stede M (2011) Lexicon-based methods for sentiment analysis. Comput. Linguist. 37:267–307

Turney PD (2002) Thumbs up or thumbs down?: semantic orientation applied to unsupervised classification of reviews. In: Proceedings of the 40th annual meeting on association for computational linguistics. Association for Computational Linguistics, pp 417–424

Turney PD, Littman ML (2003) Measuring praise and criticism: inference of semantic orientation from association. ACM Trans Inf Syst 21:315–346. doi:10.1145/944012.944013

Yi J, Nasukawa T, Bunescu R, Niblack W (2003) Sentiment analyzer: extracting sentiments about a given topic using natural language processing techniques. In: In IEEE International Conference on Data Mining (ICDM, pp 427–434)

Chapter 3
SenticNet

Abstract The main focus of this chapter lies in investigating and analysing SenticNet. The methodology by which the SenticNet sentiment lexicon was compiled has been the subject of several academic papers published by the researcher who developed SenticNet. However, little technical documentation of what is "under the hood" of SenticNet is publicly available. With this in mind and the fact that the research question posed in this paper uses SenticNet as the platform on which the research is performed, an in-depth investigation and analysis of SenticNet was carried out. The investigation consisted of a review of SenticNet-related academic papers and of an analysis of SenticNet using descriptive statistics. The goal of the investigation was to evaluate both how SenticNet contributes to concept-based sentiment analysis in terms the origins of the sentiment information contained therein, and how SenticNet could be integrated into the research. The chapter consists of seven main sections. Section 3.1 describes the sources of the knowledge in SenticNet, with particular focus on the novel aspect of the knowledge extracted from these sources. Section 3.2 describes a seminal example of sentiment analysis research. Section 3.3 presents an overview of the techniques and methods used for producing SenticNet. Section 3.4 briefly describes the core processes involved in producing SenticNet. Section 3.5 describes how RDF/XML is used to encode knowledge in SenticNet. Section 3.6 describes how this knowledge can be accessed and retrieved. Section 3.7 presents a descriptive statistics analysis of SenticNet.

Keywords AffectNet · Directed graph · Common sense knowledge · ConceptNet · RDF/XML · Hourglass of Emotions · Sentic vector

This main focus of this chapter lies in investigating and analysing SenticNet. This chapter has been produced using both primary and secondary research. Section 3.1 describes the sources of the knowledge in SenticNet, with particular focus on the novel aspect of the knowledge extracted from these sources. Section 3.2 describes a seminal example of sentiment analysis research.

© The Author(s) 2016   17
R. Biagioni, *The SenticNet Sentiment Lexicon: Exploring Semantic Richness in Multi-Word Concepts*, SpringerBriefs in Cognitive Computation 4,
DOI 10.1007/978-3-319-38971-4_3

Section 3.3 presents an overview of the techniques and methods used for producing SenticNet. Section 3.4 describes an opinion mining system in which SenticNet processes have been applied. Section 3.5 describes how knowledge in SenticNet is encoded. Section 3.6 describes how this knowledge can be accessed and retrieved. Section 3.7 presents a descriptive statistics analysis of SenticNet.

3.1 The Common Sense Nature of SenticNet Knowledge

Concepts in SenticNet are sourced from ConceptNet (Liu and Singh 2004). ConceptNet is a knowledge-representation project of the Massachusetts Institute of Technology (MIT) Media Lab.[1] The project aims to capture general human knowledge in a semantic network, in order to supply unstructured human knowledge in a structured, machine-processable form to computers. The semantic network is implemented in a directed graph data structure. The nodes of the graph are concepts, and the edges are assertions concerning the concepts. Assertions state how concepts relate to each other. For example, an assertion could be that "an oven is used for cooking". The concepts in ConceptNet are sourced from the Open Mind Common Sense (OMCS) corpus.

In 2000, the OMCS website (Singh et al. 2002) was built to collect common sense knowledge—simple assertions, descriptions of typical situations, stories describing ordinary activities and actions, and so forth—through a collaborative internet-based project. Volunteer contributors accessed the website to enter sentences in a fill-in-the blank fashion, (e.g. "The effect of eating food is _____"; "A knife is used for _____"). The website has gathered over 700,000 sentences of common sense knowledge from over 14,000 contributors around the world. The OMCS corpus consists of a diverse range of different types of common sense knowledge, expressed in natural language. The OMCS sentences alone are not directly computable (Liu and Singh 2004). The knowledge contained in the OMCS corpus includes knowledge regarding "people's common affective attitudes (e.g. desires and goals) toward situations, things, people, and actions". For example, "the last thing you do when you cook dinner is wash your dishes" or "making a mistake causes embarrassment" (Liu et al. 2003).

Several versions of ConceptNet have been developed since the project was launched in 2004. The first version of ConceptNet (Liu and Singh 2004) represented knowledge exclusively sourced from OMCS. The current version, ConceptNet 5 (Speer and Havasi 2012), has been enriched with knowledge from additional sources such as DBpedia and WordNet. DBpedia is a collaborative internet-based project to extract structured information from Wikipedia and make this information available on the Web.

---

[1]http://conceptnet5.media.mit.edu/.

3.2 A Seminal Approach to Concept-Based Sentiment Analysis

The concept-based approach to sentiment analysis has its origins in the seminal work of (Liu et al. 2003). The authors used affective common sense knowledge from the OMCS corpus to classify natural language text into one of the six "basic emotions" categories 'happy', 'sad', 'anger', 'fear', 'disgust', and 'surprise'. The output of Liu et al.'s work was a set of common sense "affect models" that could be used when classifying new text.

Many aspects of the models developed by Liu et al. are relevant to this paper's research question. These aspects, and the reasons why they are relevant, are highlighted below:

- Common sense knowledge was sourced from the OMCS corpus. From this corpus, a subset of sentences containing affective dimensions was extracted. This aspect of Liu et al.'s work is relevant for two reasons: the first is that SenticNet entries are based on knowledge sourced from OMCS; the second is that SenticNet lexicon entries must have an affective dimension in order to be included.
- The solution to the problem of classifying text into one of six "basic emotions" categories included a concept-based approach, whereby the affect of text was modelled as a function of the affect of the concepts in the text. In turn, in order to determine the affect of those concepts, the models relied on affect cues (i.e. affect keywords) found in the sentence to which the concept belonged. The result of this sequence of events is the association of concepts with one or more of the six basic emotions. These aspects of Liu et al.'s work are relevant because they represent key principles of sentiment analysis in general, and of concept-based sentiment analysis in particular. Firstly, sentiment analysis relies on "affect cues" for determining sentiment; secondly, concept-based sentiment analysis makes the implicit sentiment explicit. To illustrate, consider the sentence "Car accidents can be scary". Firstly, the affect cue, "scary", would be identified. That cue would subsequently be mapped to the basic emotion "fear", and, lastly, the concept "car accident" would be associated with "fear", effectively making the implicit sentiment of the concept explicit.
- The mechanism for mapping an affect cue to basic emotions described above uses affect keywords that are "stable across different contexts"; i.e. affect words that express the same sentiment regardless of the context in which they are used.

To summarise, Liu et al.'s seminal research exemplifies the core principles of concept-based sentiment analysis, namely that sentiment is determined by resorting to common sense knowledge, that this knowledge must have an affective dimension, and that implicit sentiment information embedded in concepts can be leveraged by making it explicit. Furthermore, Liu et al.'s work illustrates the need for affect cues (explicit or implicit) in sentiment analysis. Lastly, Liu et al.'s work echoes the research question posed in this paper by highlighting the potential context-sensitivity issue associated with single-word concepts.

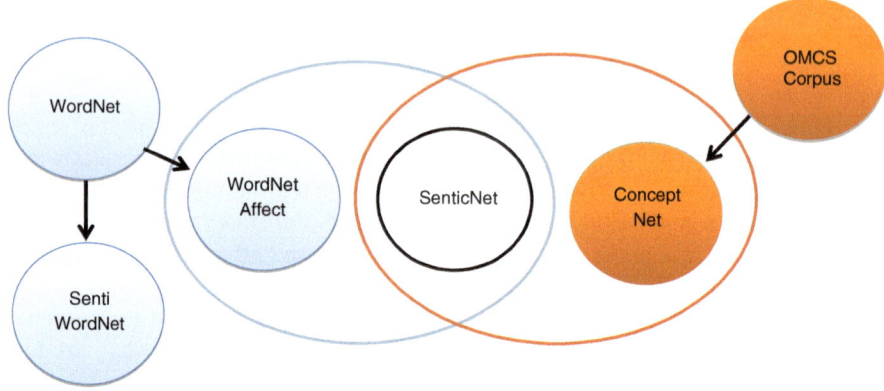

Fig. 3.1 The sentiment lexicon landscape (The Author)

3.3 Producing SenticNet

SenticNet can be considered a direct descendant of the work of Liu et al. described in Sect. 3.2. The three principles of the work of Liu et al.—namely, the sourcing of common sense knowledge from the OMCS corpus, the requirement that the knowledge must have an affective dimension and the identification of implicit sentiment through induction—are integral parts of the process used to produce SenticNet. The methods and tools employed in producing the SenticNet process are, however, different. SenticNet is the product of merging ConceptNet and WordNet-Affect (Fig. 3.1). The process by which SenticNet has been produced can be divided into two parts. The first part of the process deals with compiling the concepts stored therein. The second part determines the set of values, including polarity, associated with each concept.

In the first part, SenticNet's single-, and multi-word concepts are extracted from ConceptNet, and are merged with affective information sourced from WordNet-Affect. The process of merging the graph ConceptNet and the lexicon WordNet-Affect is achieved via two steps. In the first step, each resource is converted into a matrix. The matrix representation of ConceptNet is called AnalogySpace (Speer et al. 2008). In the second step, both matrices are combined. The resulting matrix is an "affective semantic network" named "AffectNet". AffectNet rows are concepts (e.g. 'dog' or 'bake cake'), whose columns are either common sense and affective features (e.g. 'isA-pet' or 'hasEmotion-joy'), and whose values indicate truth values of assertions. Therefore, in AffectNet, each concept is represented by a vector in the space of possible features whose values are positive for features that produce an assertion of positive valence (e.g. 'a penguin is a bird'), negative for features that produce an assertion of negative valence (e.g. 'a penguin cannot fly') and zero when nothing is known about the assertion (Cambria et al. 2011b). AffectNet effectively maps common sense knowledge (e.g. "have breakfast",

Table 3.1 Affective information and sentic vector dimensions (Cambria et al. 2010b)

	Pleasantness	Attention	Sensitivity	Aptitude
3	Ecstatsy	Vigilance	Rage	Admiration
2	Joy	Anticipation	Anger	Trust
1	Serenity	Interest	Annoyance	Acceptance
0				
−1	Pensiveness	Distraction	Apprehension	Boredom
−2	Sadness	Surprise	Fear	Disgust
−3	Grief	Amazement	Terror	Loathing

"meet people", "watch TV") to affective knowledge; i.e. WordNet-Affect's affective labels presented in Table 2.2 (Cambria and Hussain 2012, p. 38).

The degree of similarity between two concepts is the dot product between their rows in AffectNet. The value of such a dot product increases whenever two concepts are described with the same feature and decreases when they are described by features that are negations of each other. When performed on AffectNet, however, these dot products have very high dimensionality and are difficult to work with. In order to approximate them in a useful way, all of the concepts are projected from the space of features into a space with fewer dimensions i.e. the dimensionality of AffectNet is reduced. This is achieved by performing truncated singular value decomposition (TSVD) on AffectNet, resulting in a new matrix, AffectNet*, which forms a low-rank approximation of the original data. Further processing of this matrix leads to AffectiveSpace, a 100-dimensional space in which different vectors represent different ways of making binary distinctions among concepts and emotions. In AffectiveSpace common sense and affective knowledge are in fact combined, not just concomitant, i.e. everyday life concepts like 'have breakfast', 'meet people' or 'watch tv' are linked to affective domain labels. In AffectiveSpace, concepts with the same affective valence are likely to have similar features i.e. concepts concerning the same opinion tend to fall near each other in the vector space (Cambria et al. 2011b).

The second part of the SenticNet genesis process involves a variety of machine-learning techniques grouped under the umbrella term "sentic computing". Sentic computing is a novel, multi-disciplinary approach to sentiment analysis that exploits both computer and social sciences to better recognise and process natural language opinions and sentiments. In this part of the process, each SenticNet concept is associated with a "sentic vector". A sentic vector is a four-dimensional vector that can potentially express any human emotion in terms of Pleasantness, Attention, Sensitivity and Aptitude to express the affective information associated with text (Cambria and Hussain 2012, p. 51). Sentic vectors are a vector space representation of the affective common-sense knowledge derived from AffectNet, discussed above. A concept's affective information is determined from its position in the vector space. Examples of affective information associated with sentic vector dimensions are shown in Table 3.1.

The Hourglass of Emotions

The Hourglass of Emotions (Fig. 3.2) is an emotion categorisation model. The model constitutes an attempt to emulate the conception of emotions proposed in Minsky (2007), in which the human mind is perceived as a collection of resources, and emotional states are viewed as resulting from turning some set of these resources on and turning another set off. Each such selection changes how a person thinks, by changing his or her brain's activities. The state of anger, for example, appears to select a set of resources that help a person react with more speed and strength, whilst also suppressing some other resources that make the person act prudently. The Hourglass of Emotions is specifically designed to recognise, understand and express emotions in the context of human-computer interaction (HCI). In the model, in fact, affective states are not classified, as often happens in the field of emotion analysis, into basic emotional categories, but rather into four concomitant but independent dimensions—pleasantness, attention, sensitivity and aptitude—in order to understand how much the user is, respectively:

Fig. 3.2 The Hourglass of Emotion (with kind permission from Springer Science and Business Media, LCNS 7403, The Hourglass of Emotions, 2012, Cambria et al.)

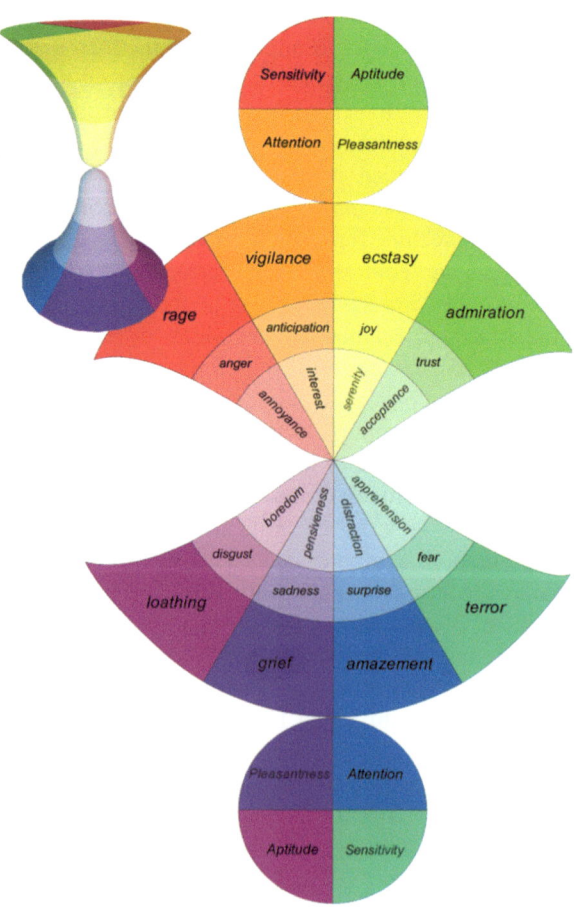

1. amused by interaction modalities (pleasantness).
2. interested in interaction contents (attention).
3. comfortable with interaction dynamics (sensitivity).
4. confident in interaction benefits (aptitude).

Each affective dimension is characterised by six levels of activation, called "sentic levels", which determine the intensity of the expressed/perceived emotion. The concomitance of the different affective dimensions makes possible the generation of compound emotions such as "love", which is given by the sum of "joy" and "trust", or "aggressiveness", given by the concomitance of "anger" and "anticipation".

AffectiveSpace and the Hourglass of Emotions
AffectiveSpace, described above, is divided into clusters using a k-NN approach according to the Hourglass model. In particular, the centroid of each cluster has been taken as the concept corresponding to each sentic level of the Hourglass model as they specify the emotional categories AffectiveSpace is organised into. As the Hourglass affective dimensions are independent but concomitant, AffectiveSpace is clustered four times, once for each dimension. According to the Hourglass categorization model, in fact, each concept can convey, at the same time, more than one emotion and this information can be expressed via a sentic vector specifying the concept's affective valence in terms of Pleasantness, Attention, Sensitivity and Aptitude (Cambria et al. 2011b). Polarity values in SenticNet are computed from the sentic vectors (Eq. 3.1).

$$p = \sum_{i=1}^{N} \frac{Pleasentness(c_i) + |Attention(c_i)| - |Sensitivity(c_i)| + Aptitude(c_i)}{3N}$$

Equation 3.1 Computing polarity from sentic vector values (Cambria et al. 2015) (with permission from Elsevier)

3.4 SenticNet Processes

This section describes the core tasks involved in producing SenticNet:

- Pre-processing tasks consisting of three tasks: firstly, affective valence indicators such as special punctuation, complete upper-case words and exclamation words are identified. These indicators are commonly found in opinionated text. Secondly, negation is detected. Lastly, the module converts text to lower case and splits the opinion into single clauses according to grammatical conjunctions and punctuation.
- The task of semantic parsing in order to deconstruct text into concepts using a lexicon of multi-word concepts. For each clause a "small bag of concepts" (SBoC).
- The task of inferring affective information; i.e. "sentics" associated with the input text. The concepts retrieved by the sentic parser are given as input to AffectiveSpace module. The concepts of each SBoC are projected into

AffectiveSpace and, according to their position, are assigned their respective affective information. The categorisation does not consist of simply labelling each concept with affect information, but also of assigning a confidence score to each label. The affective information determined by AffectiveSpace can be exploited to calculate a polarity value associated with each SBoC, as well as to detect the overall polarity associated with opinionated text.

3.5 SenticNet Knowledge: Encoding

In order to represent SenticNet knowledge in a Semantic Web aware format (Cambria et al. 2011a), its entries have been stored using a framework named RDF (Resource Description Framework) using a syntax named XML (Extensible Markup Language). The Semantic Web is a group of methods and technologies that allow machines to understand and process data based on the meaning of the data. The Semantic Web is simply a web of data described and linked in ways to establish context or semantics that adhere to defined grammar and language constructs (Dean et al. 2009). These Semantic Web methods and technologies used by SenticNet are described in this section.

RDF

SenticNet concepts and their polarity scores are encoded using the RDF data model framework developed by the W3C (World Wide Web Consortium). The purpose of RDF is to provide a standardised means of data interchange on the internet. Encoding SenticNet using RDF therefore allows its entries to be "machine-accessible and machine-processable" (Cambria and Hussain 2012, p. 65).

RDF Triples

Within the RDF framework, RDF triples are the basic unit of information in the RDF data model. RDF triples express facts via three-part statements. Triple statements consist of the elements "subject", "predicate" and "object". In an RDF triple, the subject represents a resource of interest, while a predicate is an attribute of the subject and an object is the attribute's value. SenticNet data is encoded in RDF triples. In SenticNet terms, a concept is the subject, "hasPolarity" is the predicate, and the polarity value is the object. For example, the polarity of a given concept is represented in the form "concept-hasPolarity-polarityValue". RDF triples are represented in the form of a directed graph of interconnected nodes. The graph data structure that underlies RDF triples allows for the representation of knowledge in SenticNet as a semantic network "where the semantic relatedness of concepts is retained" (Cambria and Hussain 2012, p. 83).

RDF/XML

SenticNet triples are represented in the RDF/XML format. RDF/XML is a serialisation format that provides a means of documenting an RDF model in a text-based format (Powers 2003; Segaran et al. 2009). Serialisation converts an object (in this case, the SenticNet semantic network) into a persistent form. In other words, the

RDF/XML format compacts the SenticNet semantic network into a flat file whilst preserving its graph properties.

3.6 SenticNet Access Methods

In order to perform the research, it was necessary to establish the ways in which SenticNet data could be accessed. This information would also serve to assess any potential risks encountered with any of the access methods, in terms of the availability of the resource throughout the writing of this paper. SenticNet can be accessed in three ways:

1. over the Internet via an API (Application Programming Interface)[2]
2. over the Internet via the Python programming language[3]
3. using a locally stored SenticNet RDF/XML file[4]

Access via API
SenticNet can be accessed over the Internet using an API. The syntax for the API method that returns the polarity value of a given concept is:

   http://sentic.net/api/en/concept/CONCEPT_NAME/polarity

The response of the API for the multi-word concept "accomplish goal" is shown below:

```
▼<rdf:RDF xmlns:rdf="http://www.w3.org/1999/02/22-rdf-syntax-ns#">
  ▼<rdf:Description rdf:about="http://sentic.net/api/en/concept/accomplish_goal/polarity">
    <rdf:type rdf:resource="http://sentic.net/api/concept/polarity"/>
    <polarity xmlns="http://sentic.net/api" rdf:datatype="http://www.w3.org/2001/XMLSchema#float">0.215</polarity>
  </rdf:Description>
</rdf:RDF>
```

Access via Internet and Python
SenticNet can be accessed over the Internet using the Python programming language. This access method requires using a SenticNet-specific Python package. A Python package is a collection of Python modules that implement a set of functions. The SenticNet Python package includes a function for retrieving concept polarity values.

Access via a Locally Stored RDF/XML File
SenticNet can be accessed using a locally stored copy of the SenticNet RDF/XML file, as described in Sect. 3.5. The SenticNet RDF/XML file is a persistent representation of the SenticNet semantic network. In order to query the file to retrieve a given SenticNet concept and its corresponding polarity value, the RDF triple nature of the RDF/XML file must be exploited.

---

[2]http://sentic.net/api/.

[3]https://pypi.python.org/pypi/senticnet.

[4]http://sentic.net/senticnet-3.0.zip.

3.7 SenticNet in Numbers

This section details the primary research which was performed in order to gather insights that could inform decision making for performing the research. In order to this, so an analysis of SenticNet was undertaken using the descriptive statistics methodology introduced in Sect. 1.6.

The choice of the specific descriptive statistics techniques was informed by the research question in the following ways:

- the distinction between single- and two-word concepts in the research question
- the requirement for an independent variable to be used in a controlled environment that allows the imposition of increasingly stringent classification requirements

Based on these requirements, the statistics gathered consisted of:

- an analysis of the frequency of occurrence of concepts based on the number of words they are composed of.
- an analysis of polarity values of single- and two-word concepts.

In addition, a linguistic analysis of the part-of-speech classes of SenticNet concepts was performed. The part-of-speech analysis was performed in order provide linguistic information about the concepts stored in SenticNet. While not directly related to the research question, linguistic information was deemed potentially useful for the evaluation phase of the research. The descriptive statistics employed include occurrence counts, frequency distribution, and mean and standard deviation. The analysis was performed on the most recent version of SenticNet, SenticNet 3.0 beta. The data was retrieved using a bespoke Python script and saved to a file for analysis using Microsoft Excel. The part-of-speech classes of words were determined as part of the bespoke Python script using the Python NLTK toolkit.

3.7.1 Concept Types: Number of Words

Table 3.2 presents the composition of the concepts stored in SenticNet, in terms of the number of words by concept type (i.e. whether a concept consists of one, two, or more than two words) and by polarity. The analysis of the composition is described below. Differences in summed totals are due to rounding. SenticNet contains 13,741 concepts, of which:

- 45 % (6,122) are single-word concepts, 50 % (6,839) are two-word concepts, and 6 % (780) are concepts of more than two words. If the latter concept type is excluded, then of the remaining 12,961 concepts, 47 % (6,122) are single-word concepts and 53 % (6,839) are two-word concepts.
  - Of the 6,122 single-word concepts, 59 % (3,642) have a positive polarity value, and 41 % (2,480) have a negative polarity value
  - Of the 6,839 two-word concepts, 55 % (3,749) have a positive polarity value, and 45 % (3,090) have a negative polarity value

Table 3.2 Concepts and
number of words

Number of words	Concepts with positive polarity value	Concepts with negative polarity value	Total
Single-word concept	3,642	2,480	6,122
Two-word concept	3,749	3,090	6,839
More than two words concepts	375	405	780
Total	7,766	5,975	13,741

- Of the 780 more-than-two-words concepts, 48 % (375) have a positive polarity value, and 52 % (405) have a negative polarity value
- 57 % (7,776) have a positive polarity value, and 43 % (5,975) have a negative polarity value
  - Of the 7,776 positive polarity values, 47 % (3,642) are single-word concepts, and 48 % (3,749) are two-word concepts, and 5 % (375) are concepts composed of more than two words
  - Of the 5,975 positive polarity values, 42 % (2,480) are single-word concepts, and 52 % (3,090) are two-word concepts, and 7 % (405) are concepts composed of more than two words

3.7.2 Analysis of Polarity Values: Single-Word Versus Multi-word Concepts

The results of the analysis of the frequency distributions of single- and two-word concepts, shown in Figs. 3.3 and 3.4 revealed a bell-shaped nature of the distribution of polarity values. This observation indicated that SenticNet polarity values are approximately normally distributed. This distribution allowed the author to determine the mean polarity value and standard deviations for each concept type shown in Table 3.3. Using the well-known "empirical rule" from the statistical

Fig. 3.3 Frequency distribution of polarity values for single-word concepts

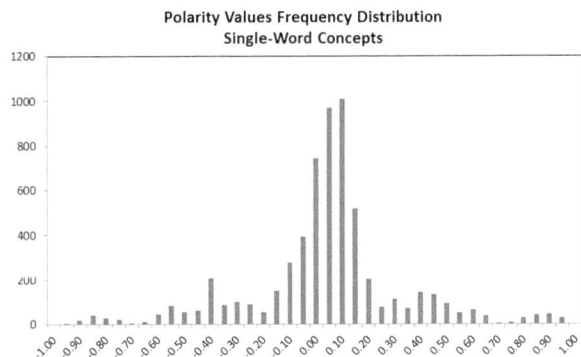

Fig. 3.4 Frequency
distribution of polarity values
for two-word concepts

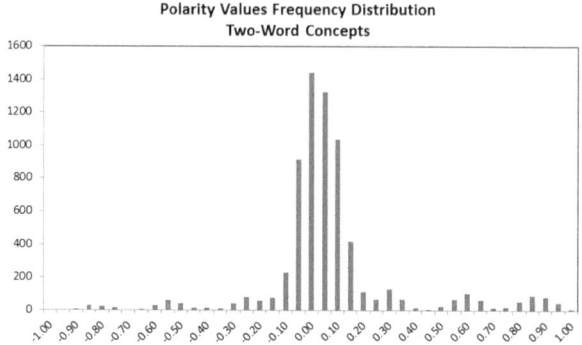

Table 3.3 Mean and
standard deviations of
polarity values

	Mean polarity value	1 standard deviation	2 standard deviations
Single-word concepts	0.017	0.289	0.577
Two-word concepts	0.047	0.257	0.514

Table 3.4 SenticNet polarity
value ranges by standard
deviations

Standard deviation(s)	Concept type	Lower bound	Upper bound	% of total number of concepts
1	Single-word concepts	−0.289	0.289	68
1	Multi-word concepts	−0.257	0.257	68
2	Single-word concepts	−0.577	0.577	95
2	Multi-word concepts	−0.514	0.514	95

sciences, an analysis was performed to determine the number of concepts that
were within one and two standard deviations from the mean polarity value. The
empirical rule of thumb expresses the heuristic that, in a normally distributed
population, approximately 99 % of values lie within three standard deviations of
the mean, approximately 95 % lie within two standard deviations, and approximately 68 % of values lie within one standard deviation (Spellman and Whiting
2013).

Applying the empirical rule to the SenticNet polarity value data allowed the
author to determine the polarity value "cut-off" points for one and two standard
deviations, displayed in Table 3.4. Those values allowed the author to determine,
for each concept type, the lower and upper bounds for isolating 68 and 95 % of

Table 3.5 Count of concepts by concept type and standard deviation

Standard deviation(s)	Single-word concepts	Two-word concepts	Total
None	6,122	6,839	12,961
1	1,580	1,119	2,699
2	420	721	1,141

concepts; i.e. concepts whose polarity values lay within one and two standard deviations of the mean polarity value, respectively.

Based on the requirements emerging from the research question discussed at the beginning of this section, and the knowledge that the higher a polarity value is, the more reliable it is (Cambria and Hussain 2012, p. 65), descriptive statistics were produced to indicate the number of concepts that would remain after excluding 68 and 95 % of concepts (Table 3.5). Of the 12,961 single- and two-word concepts in SenticNet:

- 21 % (2,699) of SenticNet concepts remain after excluding those within one standard deviation of the polarity value mean
  - of these 12 % (1,580) are single-word concepts and 9 % (1,119) are two-word concepts
- 9 % (1,141) of SenticNet concepts remain after excluding those within two standard deviations of the polarity value mean
  - of these 3 % (420) are single-word concepts and 6 % (721) are two-word concepts

3.7.3 Most common Part-of-Speech Classes

Table 3.6 illustrates that 72 % (9,329) of SenticNet's concepts, are composed of one or more nouns. This summary statistic confirmed the concept nature of SenticNet entries.

Table 3.6 Part-of-speech classes in SenticNet (Top 5)

Part-of-speech	Concepts	Cumulative (%)
Noun	3,859	30
Noun_noun	2,323	48
Verb_noun	1,819	62
Adjective_noun	1,328	72
Adjective	854	79

3.8 Conclusion

The investigation into and analysis of SenticNet as well as the review of literature specific to SenticNet, has resulted in a series of important findings. Firstly, the
work of Liu et al. supports the premise of SenticNet that affective common sense
knowledge can be leveraged to make implicit sentiment information explicit.

Secondly, the investigation into the nature of the knowledge contained in
SenticNet, and the sources from which this knowledge has been extracted indicates
that SenticNet has the potential to overcome limitations of sentiment lexicons discussed in Sect. 2.5. Specifically, the broad nature of the common sense knowledge
stored in SenticNet should overcome the issue of loss of precision associated with
the portability characteristic of domain-independent lexicons. Multi-word concepts
in SenticNet, by virtue of their inherent semantic richness, carry context information
which should reduce the loss of precision associated with portability, which in turn
should translate into a better classification performance.

Lastly, the description of SenticNet processes provides several insights into likely
requirements for the implementation of unsupervised classification. These include
the need to pre-process text (e.g. punctuation, case, grammatical structure), to handle negation (already identified as requirement in Sect. 2.2), and to extract a list of
concepts (or bag of concepts) from text and to process each concept individually. In
addition the notion of confidence scores mirrors the confidence of polarity values in
SenticNet described in Sect. 3.7.1. Furthermore, the descriptive statistics compiled
as part of the investigation provide additional input to the design and implementation
phases of the unsupervised classification procedure discussed in the next chapter.

References

Cambria E, Grassi M, Hussain A, Havasi C (2011a) Sentic computing for social media marketing.
    Multimed Tools Appl 59:557–577. doi:10.1007/s11042-011-0815-0
Cambria E, Mazzocco T, Hussain A, Eckl C (2011b) Sentic medoids: organizing affective com-
    mon sense knowledge in a multi-dimensional vector space. In: Advances in neural networks–
    ISNN 2011. Springer, pp 601–610
Cambria E, Hussain A (2012) Sentic computing: techniques, tools, and applications. Springer,
    Dordrecht. ISBN 978-94-007-5069-2
Cambria E, Gastaldo P, Bisio F, Zunino R (2015) An ELM-based model for affective analogical
    reasoning. Neurocomputing 149:443–455. doi:10.1016/j.neucom.2014.01.064
Dean M, Hebeler J, Fisher M, Blace R, Perez-Lopez A (2009) Semantic web programming,
    1st edn. Wiley, Indianapolis
Liu H, Singh P (2004) ConceptNet: a practical commonsense reasoning Toolkit. BT Technol J
    22:211–226
Liu H, Lieberman H, Selker T (2003) A model of textual affect sensing using real-world knowl-
    edge. In: Proceedings of the 8th international conference on intelligent user interfaces, IUI
    '03. ACM, New York, NY, USA, pp 125–132. doi:10.1145/604045.604067
Minsky M (2007) The emotion machine: commonsense thinking, artificial intelligence, and the
    future of the human mind. Simon & Schuster
Powers S (2003) Practical RDF, 1st edn. O'Reilly Media, Beijing , Sebastopol

Segaran T, Evans C, Taylor J (2009) Programming the semantic web, 1st edn. O'Reilly Media, Beijing, Sebastopol

Singh P, Lin T, Mueller ET, Lim G, Perkins T, Zhu WL (2002) Open mind common sense: knowledge acquisition from the general public. In: Proceedings of the first international conference on ontologies, databases, and applications of semantics for large scale information systems, lecture notes in computer science. Springer, Berlin

Speer R, Havasi C (2012) Representing general relational knowledge in ConceptNet 5. In: Proceedings of eighth international conference on language resource evaluation, LREC-2012

Speer R, Havasi C, Lieberman H (2008) AnalogySpace: reducing the dimensionality of common sense knowledge. Proceedings of the 23rd national conference on artificial intelligence—volume 1, AAAI'08. AAAI Press, Chicago, pp 548–553

Spellman FR, Whiting NE (2013) Handbook of mathematics and statistics for the environment. CRC Press, Boca Raton

Chapter 4
Unsupervised Sentiment Classification

Abstract This chapter describes the design and implementation of the unsupervised sentiment classification procedure. The classification procedure consisted of two core components: a bespoke sentiment analysis system developed by the author and the SenticNet sentiment lexicon. The sentiment lexicon acted as the source of sentiment information, and the sentiment analysis system performed the sentiment polarity classification task. The input to the sentiment classification procedure consisted of two pre-labelled datasets from two different domains. The first dataset contained sentences pertaining to the emotions "joy" and "anger". The second dataset consisted of movie reviews. The chapter contains three main sections. Section 4.1 describes the datasets used in the experiment. Section 4.2 describes the classification design and implementation. Section 4.3 describes in detail the bespoke sentiment analysis system developed in order to perform unsupervised sentiment classification.

Keywords Classification process · Knowledge encoding · Sentiment classification · RDFLib · Pickled graph · NetworkX · SPARQL

4.1 Datasets

For the purposes of performing unsupervised sentiment classification, two pairs of pre-labelled datasets from two different domains were sourced. The first dataset[1] was emotion-related. This dataset contained sentences pertaining to the emotions "joy" and "anger". The second dataset[2] contained positive and negative movie reviews. The emotion-related dataset was sourced from the ISEAR (International Survey on Emotion Antecedents and Reactions) project. The ISEAR project was run by a group of international psychologists and consisted of a survey in which respondents were asked to report situations in which they had experienced emotions of joy, fear, anger, sadness, disgust, shame, and guilt. For each emotion, the survey questions covered the

---

[1] http://emotion-research.net/toolbox/toolboxdatabase.2006-10-13.2581092615.

[2] http://www.cs.cornell.edu/people/pabo/movie-review-data/rt-polaritydata.tar.gz.

© The Author(s) 2016

R. Biagioni, *The SenticNet Sentiment Lexicon: Exploring Semantic Richness in Multi-Word Concepts*, SpringerBriefs in Cognitive Computation 4,
DOI 10.1007/978-3-319-38971-4_4

33

Table 4.1 Dataset statistics

Dataset	Polarity	Document row count
ISEAR	Positive (joy)	1079
ISEAR	Negative (anger)	1069
MOVIE	Positive	5331
MOVIE	Negative	5331

way the respondents had appraised the situation and how they reacted. Two files were created from the original ISEAR dataset. One file contained survey responses given to describe a situation associated with the emotion of "joy" which was considered to express positive sentiment. The second file was based on the emotion of "anger" which was considered to express negative sentiment. The ISEAR dataset has been used in past sentiment analysis research by Poria et al. (2012) and, again, by Poria et al. (2014b). The movie dataset, consisted of movie reviews pre-labelled as positive and negative. The dataset was made publicly available as part of past sentiment analysis research (Pang and Lee 2005). Dataset statistics are presented in Table 4.1.

The dataset labels were applied at document level. It was assumed that the sentiment polarity of all sentences within each dataset file, individually or as part of several sentences in a document row, matched the sentiment polarity of the parent file. This was considered a safe assumption in light of the fact that both datasets had been used in past sentiment analysis research, as outlined above, in which sentiment polarity classification was the research problem.

4.2 Classification Design and Implementation

4.2.1 Overview

The unsupervised sentiment classification (Fig. 4.1) was loosely modelled on past sentiment analysis research by Weichselbraun et al. (2013) and Poria et al.

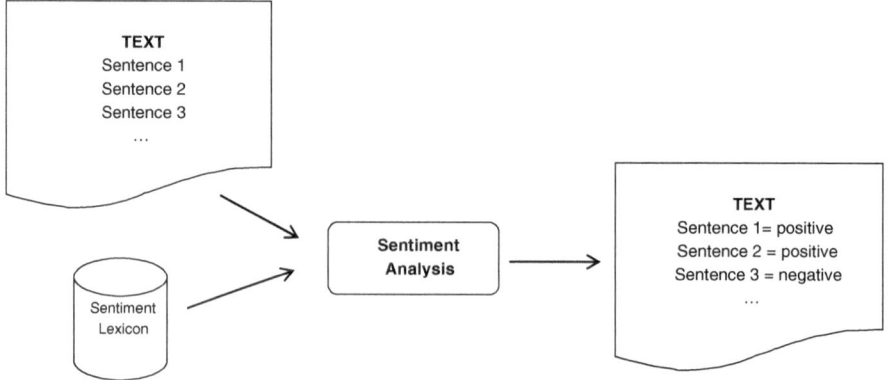

Fig. 4.1 High-level classification design (The Author)

(2014a). The classification procedure consisted of two core components: a bespoke sentiment analysis system and the SenticNet sentiment lexicon. The sentiment lexicon acted as the source of sentiment information, and the sentiment analysis system performed the sentiment polarity classification task. The input to the sentiment classification procedure consisted of the datasets discussed in Sect. 4.1. The output consisted of a text file containing the classification results. The decision to model sentiment classification on Weichselbraun et al. (2013) and Poria et al. (2014a) resulted in the adoption of the sentence-level as the level of analysis for the classification of sentiment.

4.2.2 Sentiment Classification Process

The sentiment classification process consisted of submitting each row in each dataset polarity file to the classification procedure. If a row consisted of more than one sentence, each sentence was classified separately. It was assumed, as noted in Sect. 4.1, that the sentiment polarity of each document row, whether consisting of one or more sentences, matched the sentiment polarity of the file it was extracted from.

The process had to meet two key requirements. The first requirement was to submit each sentence to two classification attempts, one by the single-word concept strategy and one by the multi-word concept strategy. The term "strategy" refers to the two different ways in which the classification procedure attempts to classify sentences. Strategies were implemented by generating two bags of concepts from each sentence, one containing single-word concepts, the other containing multi-word concepts.

The second requirement was that attempts had to return a classification score, regardless of whether the score was correct or not. This requirement was based on the need to compare results between both concept types at the sentence level without those results being affected by limitations of the SenticNet lexicon. The limitations relate to the issue of lexicon coverage, which could have introduced randomness into the results because one concept type strategy may not have found a match in SenticNet, while the other had, thus invalidating a comparison between both strategies in terms of ability to accurately determine sentiment. Therefore, if one of the two concept type strategies, or both, failed to determine a classification score, the sentence could not be included in the set of results. A sample output of the sentence classification output is shown in Table 4.2. An overview of the classification process is shown in Fig. 4.2.

4.2.3 Polarity Value Thresholds

The review of SenticNet related literature discussed in Chap. 3 highlighted that polarity values in SenticNet intrinsically express a degree of confidence in their reliability. The closer to -1 or 1 a polarity value is, the higher the confidence of

Table 4.2 Sample output of the sentence classification process

RowID	SentenceID	Score of single-word strategy	Score of multi-word strategy
1	1	0.24	0.11
2	1	0.8	−0.2
2	2	−0.65	−0.01
3	1	0.2	0.23

Table 4.3 Polarity value ranges for single-word concepts by polarity value threshold

Single-word concepts		
PVT	Negative polarity	Positive polarity
0	−1 to 0	0 to 1
1	−1 to −0.289	0.289 to 1
2	−1 to −0.577	0.577 to 1

Table 4.4 Polarity value ranges for multi-word concepts by polarity value threshold

Multi-word concepts		
PVT	Negative polarity	Positive polarity
0	−1 to 0	0 to 1
1	−1 to −0.257	0.257 to 1
2	−1 to −0.514	0.514 to 1

Table 4.5 Top 5 multi-word concepts by polarity value threshold

		Polarity		Polarity
PVT0	want_degree	0.02	wooden_spoon	−0.023
	child_play	0.023	train_track	−0.023
	chess_pawn	0.023	tool_shed	−0.023
	after_lunch	0.023	service_station	−0.023
	buy_store	0.023	view_nature	−0.023
PVT1	grow_up	0.29	use_drug	−0.29
	pay_tuition	0.291	smelly_foot	−0.29
	read_program	0.291	death_row	−0.29
	make_coffee	0.291	lack_time	−0.29
	birthday_cake	0.292	volcanic_eruption	−0.291
PVT2	nice_look	0.579	back_pain	−0.579
	religious_ceremony	0.579	break_arm	−0.58
	enough_food	0.58	mess_up	−0.581
	desire_knowledge	0.58	hurt_others	−0.581
	graduation_ceremony	0.581	rotten_food	−0.581

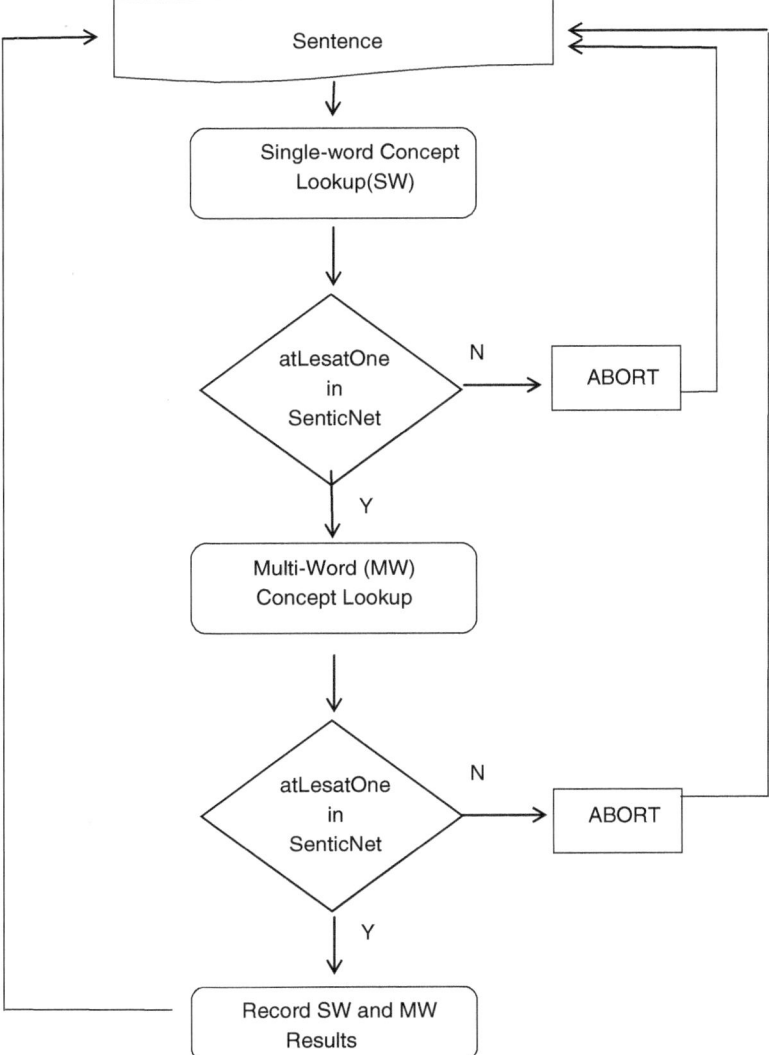

Fig. 4.2 Sentence classification process (The Author)

SenticNet that the value is representative of the true sentiment orientation. The idea of quality being associated with polarity values was seen as echoing the proposition made in this paper that semantic richness is associated to multi-word concepts. For this reason polarity value reliability was introduced as a variable into the classification design in the form of a threshold value.

The purpose of the polarity value threshold was to set a minimum polarity value magnitude that a concept had to meet in order to be eligible for inclusion into the classification process. In this way, the SenticNet concepts eligible for contributing to the classification outcome could be restricted to concepts with a high degree of confidence in their reliability. The integration of polarity value thresholds into the design was based on the "controlled experiment" model. A controlled experiment is an experiment observing the effect of changes to one variable (called an independent variable) on another variable (called a dependent variable). All other variables in a controlled experiment remain unchanged, in order to minimise the effects of variables other than the single independent variable. The goal of a controlled experiment design is to increase the reliability of results. By conceiving the classification procedure as a controlled experiment it was possible to provide a framework that allowed to provide some certainty to the assertion that an observed increase in classification accuracy, if any, was caused by increasing the reliability requirements imposed on polarity values.

The polarity threshold values (PVT) were chosen by applying a statistical framework. The framework was based on the standard deviations, described in Sect. 3.7, of the single-word and multi-word concept polarities in SenticNet. Threshold values were derived from applying one and two standard deviations respectively to the mean polarity values of each concept type. PVTs were named "PVT1" and "PVT2" depending on the number of standard deviations. PVT1 and PVT2 excluded, as described in Sect. 3.7, 68 and 95 % of polarity values with low reliability respectively. An additional threshold of zero, named "PVT0", was used as a baseline value. PVT0 imposed no minimum polarity value to SenticNet concepts. The polarity value ranges resulting from applying PVTs are shown in Tables 4.3 and 4.4. For a given threshold, the retrieved polarity value had to be smaller than the negative threshold value or greater than the positive threshold value. The qualitative impact of PVT's on the qualifying SenticNet concepts is illustrated in Table 4.5.

4.2.4 Implementation

The requirements of the bespoke sentiment analysis system were derived from the limitations imposed by SenticNet, the requirements of the research question and from the analysis of SenticNet described in Chap. 3. Requirements derived from the limitations imposed by SenticNet related to availability, i.e. SenticNet had to be available at all times, and knowledge-encoding, i.e. the way in which knowledge is represented in SenticNet (Sect. 3.6).

Requirements derived from the research question described in Sect. 1.3 related to the implementation of PVTs. System requirements derived from the study of the SenticNet processes described in Sect. 3.4 i.e. to the handling of negation, and of sentence structure. Other system requirements were imposed by limitations inherent to natural language processing (semantic integrity, word form, processing time), and sentiment analysis (negation handling). An overview of the requirements of the sentiment analysis system, and the type of solutions implemented, is provided in Table 4.6.

Table 4.6 System requirements overview

Req. ID	Requirement category	Requirement	Implementation type
1	SenticNet	Availability	RDFLib, NetworkX and SPARQL
2	SenticNet	Knowledge encoding	RDFLib, NetworkX and SPARQL
3	Concept	Semantic integrity sentence	Sentence splitting
4	Concept	Semantic integrity word order	Combinations
5	Concept	Word form agreement	Normalisation
6	Sentiment analysis	Negation handling	Custom list of negation words
7	Text processing	Minimising processing time	Stopword and punctuation removal

Fig. 4.3 SenticNet
RDF/XML lookup process
(The Author)

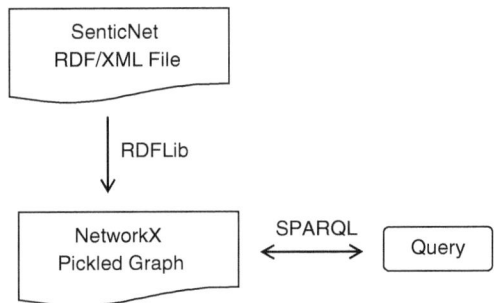

Fig. 4.4 Pseudocode negation
handling function

negation-word-list = (...)

negation-flag = 0

FOR EACH word IN sentence

   IF word in negation-word-list

      set negation-flag = 1

RETURN FLAG

Availability (Requirement ID 1)

The system had to be able to access SenticNet at all times during the time of writing this paper. To meet this requirement the SenticNet RDF/XML file, available from the SenticNet website, was downloaded and saved locally. All interactions of the system with SenticNet occurred through this file.

Knowledge Encoding (Requirement ID 2)

The program had to be able to meet the requirements of the knowledge representation schema of SenticNet discussed in Sect. 3.6). , namely the RDF data model, RDF Triples and the RDF/XML format. To meet this requirement the Python library RDFLib, the Python-based graph toolkit NetworkX and the query language SPARQL (Simple Protocol and RDF Query Language) were used to implement the knowledge-encoding aspects of the research. RDFLib is a Python library used for working with RDF. The library contains an RDF/XML parser and serialiser (serialisation was discussed in Sect. 3.6). NetworkX is a Python-based software package used for the creation and manipulation of networks. SPARQL is a query language designed to query data conforming to the RDF data model. SPARQL expresses queries using triple patterns, which is similar to the RDF triples described in Sect. 3.6).

The SenticNet lookup process (Fig. 4.3) proceeded as follows: in the first step, RDFLib parsed the SenticNet RDF/XML file. The parsing of the file allowed the author to re-create the semantic network structure of SenticNet. Subsequently, NetworkX wrote the parsed RDF/XML file to a "pickled graph" object. A "pickle" is a NetworkX module that supplies serialisation mechanism functions (DuCharme 2011). In this instance the pickled graph object's serialisation mechanisms enabled the querying of SenticNet's entries. The pickled graph object allowed the author to retrieve the polarity value of a given concept from SenticNet via a SPARQL query.

Semantic Integrity (Requirement ID 3 & 4)

The system had to be able to extract multi-word concepts from each sentence whilst also ensuring that those multi-word concepts would be faithful to the meaning of a given sentence. Automatically extracting concepts from a sentence could be achieved in two ways: employing linguistic analysis of the syntactic composition of the sentence or adopting a brute force approach. The linguistic analysis approach would have required expertise in manipulating sentence syntax using specialised tools such as the Java-based Stanford Core NLP toolkit. The brute force approach did not rely on linguistic information, but instead simply involved extracting all unigrams and bigrams from a sentence (a unigram is a single word, while a bigram is a pair of consecutive words), knowing that some of them would represent actual concepts. The brute force approach was chosen for extracting multi-word concepts because of its simplicity: generating multi-word concepts using the brute force approach could be accomplished by generating two-word combinations or permutations from all words in a given sentence. While simple, this approach carried two risks related to sentence structure and to word order.

Generating two-word combinations or permutations from all available words in a sentence, regardless of sentence structure, could potentially have led to multi-word concepts that were not part of the original meaning of a sentence. Specifically, combining words to form a multi-word concept from two different sentence clauses could have produced undesired results. The risk of creating unwanted two-word combinations or permutations was mitigated by restricting the word list from which to generate two-word combinations to the boundaries of sentence clauses. A clause is "a unit of grammatical organization", ranked next below the sentence itself, and is easily identifiable by its enclosing commas. For example, "I had a last drink for the road, and then drove home. I was lucky there was no traffic control check point on the way." consists of two sentences, with the first sentence consisting of two clauses.

As noted above, there are two possibilities for generating multi-word concepts from a list, or bag, of words: combinations or permutations. The permutation approach ignores the order in which words appear, and could potentially have led to multi-word concepts that were not part of the original meaning of a sentence. For example, in the clause "but I feel joy too because her life is taking the right direction", permutations would generate the multi-word concept "take life", which does not reflect the meaning of the sentence. However, word permutations may have led to multi-word concepts that a word combination approach could not have produced. For example, in the clause "freedom of speech was tightly restricted", permutations would have generated the multi-word concept "restrict freedom", which does reflect the meaning expressed in the clause. In order to avoid the risk of creating unwanted multi-word concepts, the combination approach was preferred over the permutation approach. The fact that opportunities would be missed by not using word permutations was accepted as a necessary trade-off for minimising error.

Word Form Agreement (Requirement ID 5)

The system had to be able to transform the word form of concepts in text such that they would be in agreement with the word form of concepts in SenticNet. Verbs in SenticNet are stored in present tense form. Nouns are stored in singular form. These facts had implications for the process of looking up concepts in SenticNet, because a mismatch between word forms would prevent a potential match from being found. Incomplete scoring due to a mismatch of word forms would have introduced error into the classification process. Verb-tense conversion was performed using the Python-based library "NodeBox Linguistics" The Nodebox toolkit allowed the author to perform grammar inflection and semantic operations on English language text, and included a part-of-speech tagging function. This function was used to test whether a given word was a verb. Plural-to-singular noun conversion was performed by the Python package "Inflect". The Inflect package generates plurals, singular nouns, ordinals, indefinite articles, etc.

Negation Handling (Requirement ID 6)

The system had to be able to handle sentiment negation. Negation refers to situations in which the sentiment orientation in a sentence is inverted by negation words; for example, "not good", or "I do not like pie". In sentiment analysis systems, negation-handling is of critical importance (Liu 2012), and is a non-trivial task, as described in Sect. 3.5. In order to accurately handle negation in text, it is necessary to perform linguistic analysis (Mutalink et al. 2001, cited in Ohana 2009, p.122). Similar to Requirement ID's 3&4, the linguistic analysis of text would have required expertise in manipulating sentence syntax using specialised tools. A simpler alternative was to perform a coarse negation test on each sentence by detecting the presence of commonly used negation words. Thus, sentiment negation was handled by developing a custom negation handling function

(Fig. 4.4). The function was based on a list of pre-defined negation words against which each word in a given sentence was tested. If a negation word was found in a sentence, the sign of the final polarity score would be inverted to reflect the sentiment negation in the final score. The negation handling function was applied at the beginning of each classification procedure. A negation flag would be set where appropriate, and the flag would be checked at the end of the process to decide whether the sign of the final polarity score had to be inverted or not.

4.3 Conclusion

The design and implementation of an unsupervised classification procedure in general, and the bespoke sentiment analysis system in particular, described in this chapter are based on the requirements emanating from the research question and from SenticNet. The research question requirements defined the nature of the input text and the strategy of the sentence-level sentiment classification procedure. SenticNet defined the techniques for identifying concepts and for retrieving their polarity values, as well as the threshold values used to submit the two single-word and multi-word classification strategies to a series of classification tasks in which the quality requirements of SenticNet were increasingly heightened. Furthermore, this chapter described how the classification procedure was modelled on past research in the same domain of concept- and lexicon-based sentiment analysis to provide a sound foundation for unsupervised sentiment classification.

The chapter also highlighted two shortcomings of the bespoke sentiment analysis system. Firstly, generating multi-word concepts from all the words in a sentence clause, whether using combinations or permutations, produced many meaningless word combinations. In this sense the efficiency of the proposed system could be improved. Second, the proposed solution to negation handling is simplistic and may impede the effective detection of negation during the classification procedure. As a whole, however, the bespoke sentiment analysis system, and the classification procedure was shown to have been designed and implemented such that classification procedure was capable of providing answers to the research question—whether multi-word concepts, by virtue of their semantic richness, outperform single-word concepts in an unsupervised sentence-level sentiment classification task which uses these two types of concepts as the sole classification feature. The answers to the research question are presented in the next chapter.

References

DuCharme B (2011) Learning SPARQL, 1st edn. O'Reilly Media, Sebastopol
Liu B (2012) Sentiment analysis and opinion mining. Synth Lect Hum Lang Technol 5:1–167. doi:10.2200/S00416ED1V01Y201204HLT016

Ohana B (2009) Opinion mining with the SentWordNet lexical resource. Dublin Institute of Technology

Pang B, Lee L (2005) Seeing stars: exploiting class relationships for sentiment categorization with respect to rating scales. In: Proceedings of the 43rd annual meeting on association for computational linguistics, ACL '05. Association for computational linguistics, Stroudsburg, PA, USA, pp 115–124. doi:10.3115/1219840.1219855

Poria S, Gelbukh A, Cambria E, Yang P, Hussain A, Durrani T (2012) Merging SenticNet and WordNet-Affect emotion lists for sentiment analysis. In: Presented at the international conference on signal processing proceedings, ICSP

Poria S, Cambria E, Winterstein G, Huang G-B (2014a) Sentic patterns: dependency-based rules for concept-level sentiment analysis. Knowl-Based Syst 69:45–63. doi:10.1016/j.knosys.2014.05.005

Poria S, Gelbukh A, Cambria E, Hussain A, Huang G-B (2014b) EmoSenticSpace: a novel framework for affective common-sense reasoning. Knowl-Based Syst 69:108–123. doi:10.1016/j.knosys.2014.06.011

Weichselbraun A, Gindl S, Scharl A (2013) Extracting and grounding contextualized sentiment lexicons. IEEE Intell Syst 28:39–46. doi:10.1109/MIS.2013.41

Chapter 5
Evaluation

Abstract This chapter analyses and presents the results of the sentiment classification procedure designed to explore the semantic richness of multi-word concepts in a context of sentiment analysis. The outcomes of the classification procedure do not support the idea that the semantically richer multi-word concepts outperform single-word concepts in an unsupervised sentiment classification task. However results suggested that, by increasing the reliability requirements on the sentiment information provided by SenticNet, a better environment is created for the semantic richness of multi-word concepts to become a contributing factor in classification accuracy. The chapter contains two sections. In Sect. 5.1 results are summarised and interpreted in light of several aspects of the classification procedure. Section 5.2 provides results directly addressing the research question. Section 5.3 discusses how qualitative differences between the datasets may have affected the results. Sections 5.4–5.6 discuss the results directly relating to the SenticNet sentiment lexicon, the bespoke sentiment analysis system and the design of the classification procedure. Section 5.7 briefly discusses the limitations of the classification procedure that may have adversely affected the results.

Keywords Classification accuracy · Polarity value · Multi-word concepts · Negation handling

5.1 Classification Performance

Classification outcomes are summarised and interpreted in this section. All outcomes in this section concern the classification accuracy achieved by the multi-word and single-word concept strategies. An overview of the results is presented below in Figs. 5.1, 5.2, 5.3 and 5.4. In these figures, the abbreviations "SW", "MW" and "PVT" stand for single-word concepts, multi-word concepts, and polarity value threshold, respectively.

© The Author(s) 2016
R. Biagioni, *The SenticNet Sentiment Lexicon: Exploring Semantic Richness in Multi-Word Concepts*, SpringerBriefs in Cognitive Computation 4,
DOI 10.1007/978-3-319-38971-4_5

Fig. 5.1 Classification
accuracy for positive ISEAR
dataset

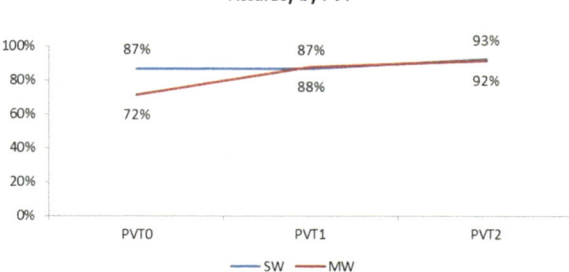

Fig. 5.2 Classification
accuracy for negative ISEAR
dataset

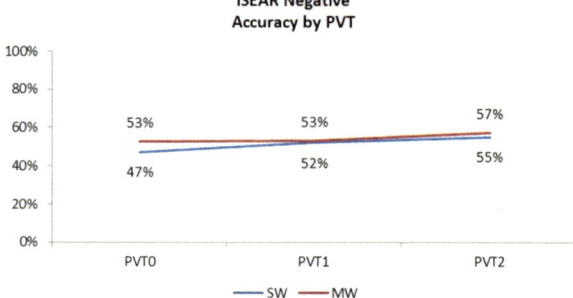

Fig. 5.3 Classification
accuracy for positive MOVIE
dataset

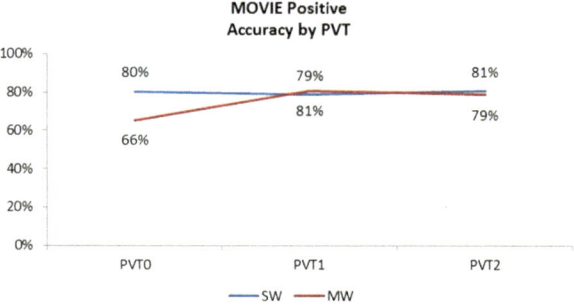

Fig. 5.4 Classification
accuracy for negative MOVIE
dataset

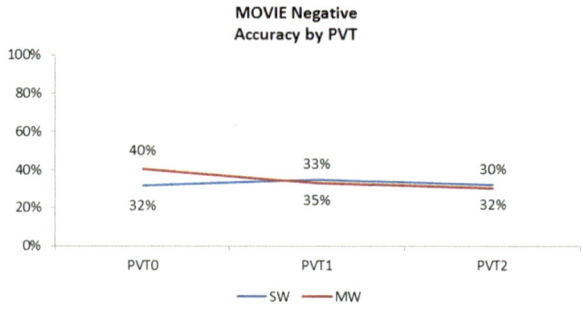

5.2 Research Question

The research results do not support the proposition, presented in Sect. 1.3, that SenticNet could gain from solely containing multi-word concepts. The reasons for this interpretation is that, firstly, the results of the classification procedure showed that multi-word concepts outperformed single-word concepts only in cases of negative polarity datasets. Specifically, where no threshold was applied, the multi-word concept strategy outperformed the single-word concept strategy by 6 % (Fig. 5.2) on the negative ISEAR dataset, and by 8 % (Fig. 5.4) on the negative MOVIE dataset. The better accuracy suggested by the results of multi-word concepts, however, was not sustained: as PVT1 and PVT2 were introduced, the accuracies of both concept types reached similar levels.

Secondly, in the case of positive polarity datasets, multi-word concepts were outperformed by single-word concepts. Specifically, where no threshold was applied, multi-word concepts were outperformed by single-word concepts by 15 % (Fig. 5.1) on the positive ISEAR dataset, and by 14 % (Fig. 5.3) on the positive MOVIE dataset. Similar to the results observed for the negative datasets, the better accuracy of the results for single-word concepts on positive datasets was not sustained: as PVT1 and PVT2 were introduced, the accuracies of both concept types reached similar levels.

Thirdly, the introduction of PVTs resulted in most cases in the differences in classification accuracies between concept types "levelling off". The effect of PVT1 on the multi-word concept type in the positive datasets was seen to be more impactful: a 16 % increase in classification accuracy of the multi-word concept strategy for the positive ISEAR dataset (Fig. 5.1) and a 15 % increase for the positive MOVIE dataset (Fig. 5.3), compared to a 5 % increase in classification accuracy of the single-word concept strategy for the negative ISEAR dataset (Fig. 5.2) and a 3 % increase for the negative MOVIE dataset (Fig. 5.4).

In light of these observations, while results do not support the idea that multi-word concepts, by virtue of their semantic richness, consistently outperform single-word concepts in an unsupervised sentiment classification task, they do suggest that, by increasing the reliability requirements on the sentiment information provided by SenticNet, a better environment is created for the semantic richness of multi-word concepts to become a contributing factor in classification accuracy.

5.3 Qualitative Differences Between the Datasets

The effect of the restrictions imposed by PVT's on the classification outcome suggested qualitative differences in the two MOVIE and ISEAR datasets. Specifically, the restrictions imposed by the scoring rule described in Sect. 4.2.2 resulted in only 21 % of sentences of the MOVIE dataset being scored, compared to 53 % of the sentences of the ISEAR dataset (Tables 5.1 and 5.2). Furthermore, on average

Table 5.1 Number of sentences scored by dataset polarity and PVT

		Total dataset	Total scored		
			PVT0	PVT1	PVT2
ISEAR	Positive	1079	561	212	120
	Negative	1069	578	175	49
MOVIE	Positive	5331	1064	241	113
	Negative	5331	1196	251	115

Table 5.2 Proportion of sentences scored by dataset polarity and PVT

		Total dataset (%)	Total scored		
			PVT0 (%)	PVT1 (%)	PVT2 (%)
ISEAR	Positive	100	52	20	11
	Negative	100	54	16	5
MOVIE	Positive	100	20	5	2
	Negative	100	22	5	2

the effect of the restrictions imposed by the PVTs reduced the number of scored sentences in the ISEAR dataset to 18 % (PVT1) and 8 % (PVT2), and to 5 % (PVT1) and 2 % (PVT2) in the MOVIE dataset (Tables 5.1 and 5.2).

The fact that the impact of PVT's was less pronounced in the ISEAR dataset than in the MOVIE dataset could have two explanations. The first explanation could be that the SenticNet lexicon is better suited to emotion-related natural language text than to language used to express opinions about movies. This would be plausible in light of the fact that SenticNet, as described in Sect. 3.1, consists of affective concepts extracted from ConceptNet. The second explanation could be that there is little agreement between the writing style and vocabulary of the MOVIE dataset, on the one hand, and that of the common sense knowledge on which SenticNet is built, on the other. While the latter tends to be factual (e.g. "a stove is used for cooking"), the former tends to be fragmented, interspersed with a large number of facts about the movie, or informal as illustrated by the following sentence extracted from the negative polarity movie dataset: "for single digits kidlets stuart little 2 is still a no brainer. if you're looking to rekindle the magic of the first film, you'll need a stronger stomach than us."

5.4 SenticNet

The evaluation of the results supported the reliability aspect of SenticNet's polarity values and SenticNet's domain independence. The notion of reliability, intrinsic to the magnitude of polarity values, was confirmed by the observation that in most cases classification accuracy increased when PVT1 was applied to the classification procedure. Furthermore, the sharp increase in accuracy of multi-word concepts observed in both datasets when PVT1 was introduced suggested that the SenticNet lexicon is domain-independent.

5.5 Sentiment Analysis System

The results demonstrated that the sentiment analysis system achieved higher classification accuracy on positive datasets (Figs. 5.1 and 5.3) than on the negative datasets (Figs. 5.2 and 5.4). This could be due to several reasons. Firstly, it is more difficult to detect negative sentiment. For example, as discussed in Sect. 2.1, negation can be expressed using sarcasm and irony. Secondly, the rudimentary negation handling function of the bespoke sentiment analysis system may have resulted in the system not recognising all cases of negation.

5.6 Sentiment Classification Design

The results of the classification procedure, in terms of the number of sentences scored, highlighted that the choice of polarity value thresholds was too extreme in the case of PVT2 (Tables 5.1 and 5.2). Thus, the results obtained for PVT2 cannot be said to be representative of the ability of single- and multi-word concepts in SenticNet to classify sentences accurately.

5.7 Limitations

The limitations of the classification procedure that may have adversely affected the outcome include:

- The simplistic negation handling algorithm implemented as part of the bespoke sentiment analysis system may have affected the degree to which negation was detected, and how accurately it was detected.
- Multi-word concepts were restricted to two words, in order to minimise the computational expense resulting from the production of word combinations from lists of words, as described in Sect. 4.3.2. This meant that the SenticNet lexicon was not exploited to its full potential, because it contains concepts that are longer than two words (e.g. "love another person"). Restricting the possible multi-word concepts that were looked up in SenticNet represents a limitation of the classification design because it potentially misses out on opportunities for multi-word concept classification. This limitation was mitigated by the fact that,

as shown in Sect. 3.7, the restriction to two-word concepts affected only 6 % of all SenticNet entries.

5.8 Conclusions

The key learnings of this chapter related to the insights gained from the classification procedure and the limitations that may have affected those results. The proposition on which the research question was formulated, namely that SenticNet would benefit from solely being an exclusively multi-word concept sentiment lexicon could not be conclusively demonstrated. Nevertheless, the perceived semantic richness of multi-word concepts has received backing from the classification results by the observation that the requirement for higher quality SenticNet sentiment information resulted in a sharp increase in the classification accuracy of the multi-word concept strategy across two datasets, from two different domains.

Additional learnings included indications of a possible inter-dependence, in terms of classification performance, between writing style in natural language text and the origins of entries in a sentiment lexicon on classification outcomes. Further, the reliability expressed by the magnitude of SenticNet polarity values was confirmed and results suggested that the lexicon is domain independent. The choice of the second threshold, PVT2, proved to be too extreme in its effect on the number of SenticNet entries it excluded which, in turn, had a detrimental effect on the number of sentences scored.

Lastly, the relatively poor accuracy of results of the negative datasets, compared to the positive datasets, suggested that the negation handling function of the bespoke sentiment analysis was too simplistic to effectively recognise negation. Limitations of the bespoke system also included the fact that only two-word concepts were considered for the multi-word concept strategy. The limited number of datasets was also highlighted as a factor limiting the ability of the results to yield conclusions with regards to the research question.

Chapter 6
Conclusion

Abstract The research presented in this paper aimed to review existing sentiment analysis approaches, techniques, and resources, and their limitations; to perform a detailed evaluation of SenticNet to understand how SenticNet was compiled, and how it is different from other similar resources. This review was then employed in the design and implementation of a classification procedure aimed at answering the research question whether multi-word concepts, by virtue of their semantic richness, outperform single-word concepts in an unsupervised sentence-level sentiment analysis task which uses these two types of concepts as the sole classification feature. This chapter presents the conclusions reached from the attempt of answering the research question, briefly discusses potential future work and concludes with final remarks.

Keywords Multi-word concepts · Knowledge encoding · Natural language · SenticNet · Semantic richness

6.1 Conclusion

The research presented in this paper aimed to review existing sentiment analysis approaches, techniques, and resources, and their limitations; to perform a detailed evaluation of SenticNet to understand how SenticNet was compiled, and how it is different from other similar resources. This review was then employed in the design and implementation of a classification procedure aimed at answering the research question whether multi-word concepts, by virtue of their semantic richness, outperform single-word concepts in an unsupervised sentence-level sentiment analysis task which uses these two types of concepts as the sole classification feature.

The research performed as part of this paper suggests that there is little discernible difference between single-word concepts and the semantically richer multi-word concepts in terms of their accuracy in classifying sentiment in unsupervised sentence-level classification, where concept type is the only classification feature.

© The Author(s) 2016
R. Biagioni, *The SenticNet Sentiment Lexicon: Exploring Semantic Richness in Multi-Word Concepts*, SpringerBriefs in Cognitive Computation 4,
DOI 10.1007/978-3-319-38971-4_6

In fact, quantifying semantic richness has proven to be an elusive goal, similar to attempting to make the intangible tangible. The suggestion made in Chap. 5 that increasing the quality requirement on SenticNet entries favours the semantic richness inherent to multi-word concepts is an impression gained from observing trend lines in graphs and could not be reinforced with a statistical test. Further, the suggestion that the polarity of natural language text, as well as writing style, impact the outcome of sentiment analysis, could not be supported by scientific evidence within the parameters of this paper. On the positive side, the author found that the RDF encoding of SenticNet knowledge provided an efficient framework for querying SenticNet. Further research is required to better analyse this finding.

6.2 Future Work

The research performed in this paper has uncovered opportunities for future research which could lead to interesting results. These opportunities are outlined below:

- Re-evaluating the objective of making semantic richness tangible by designing a similar classification procedure but with additional features to be used in the classification procedure and possibly using a supervised learning approach
- Conducting research into the effectiveness of the SenticNet RDF knowledge encoding framework, compared to other sentiment lexicons
- Conducting research into the perceived domain independence of SenticNet. In fact, in an email exchange with Cambria (founder of SenticNet), it was suggested that domain specific versions of SenticNet were being considered. However, it would be interesting to explore the degree to which the current SenticNet version is able to reliably cover different domains.
- Conducting research into the perceived impact of the polarity of natural language text on classification performance and whether any differences are due to negation handling or to writing style.

6.3 Final Remarks

There are over a trillion pages of information on the Web, almost all of it in natural language. An agent that wants to do knowledge acquisition needs to understand (at least partially) the ambiguous, messy languages that humans use (Russell and Norvig 2009, p. 860).

This paper has shown that common sense knowledge can be captured at scale through collaborative initiatives such as the Open Mind Common Sense Project, and that the semantically rich nature of concepts can be harnessed by techniques such as sentic computing. The bringing together of the potentially unlimited supply of common sense knowledge and sentic computing has the potential to transmit to computers knowledge about the world previously inaccessible to them. This could lead to developing emotionally intelligent machines that are able to grasp the way we feel, for better or for worse…

Reference

Russell S, Norvig P (2009) Artificial intelligence: a modern approach, 3rd edn. Prentice Hall, Upper Saddle River

Index

A
Affective knowledge, 13, 20, 21
AffectiveSpace, 21, 23
AffectNet, 20, 21

C
Classification accuracy, 34, 44, 45, 47, 48
Classification process, 33–35, 39
Common sense knowledge, 18–20, 29, 46, 50
ConceptNet, 18, 20, 46
Concepts, 4, 5, 14, 18–20, 23, 24, 26–30, 34, 38–40, 43–45, 47–50

E
Evaluation, 3, 26, 43, 47, 49

H
Hourglass of emotions, 21, 23

K
Knowledge encoding, 38, 50

L
Linguistics, 7, 11, 39

M
Machine learning, 9, 11

N
Natural language, 3, 4, 9, 11, 12, 14, 18, 21, 37, 46, 48–50

P
Part-of-speech, 4, 9, 11, 12, 14, 26, 29, 39
Polarity, 4, 5, 7–13, 20, 23, 25–28, 32–34, 38, 39, 44, 46–49

R
Reliability, 33, 34, 45, 47, 48

S
Semantic richness, 5, 29, 40, 45, 48–50
SenticNet, 4, 5, 7, 14, 17–19, 21, 23–25, 27–29, 32–34, 37, 39, 45–50
Sentic vector, 21, 23
Sentiment analysis, 2–4, 7–14, 17, 19, 21, 31, 32, 39, 47, 48
Sentiment classification, 4, 5, 10, 32, 33, 40, 43, 47
Sentiment lexicon, 4, 5, 7, 9, 12–14, 29, 48
Statistical analysis, 27
Statistical framework, 34
Synsets, 12, 13
Syntactic categories, 11

U
Unsupervised learning, 7, 10

© The Author(s) 2016
R. Biagioni, *The SenticNet Sentiment Lexicon: Exploring Semantic Richness in Multi-Word Concepts*, SpringerBriefs in Cognitive Computation 4,
DOI 10.1007/978-3-319-38971-4